Dedicated to the great ones....

Mr & Mrs H.U Owoh

ACKNOWLEDGEMENTS

I will like to express my heartfelt appreciation to Dr David Naseby and Dr Timothy Aldsworth, despite their busy schedule provided adequate support and guidance during the course of this book.

I will also like to express my sincere gratitude to my parents, Mr and Mrs Owoh for providing me with financial and moral support. In addition, i will like to thank my sisters Grace and Charity Owoh and to the rest of my family for their contributions. Finally, I will like to offer my regards to the special ones CGF members and not forgetting John Omizu and Oluwakemi Kayode.

Contents   Pages

Chapter 3: Identifying significant medium constituents using Plackett and Burman Method. 87

Chapter 4: One factor L-cysteine.HCl.H_2O medium optimization design   127

Figures

Pages

Tables   Pages

Chapter 1

Introduction and Literature Review

1.1 Overview of Hydrogen production

1.1.1 The need for green energy production

A great proportion of our energy requirements are provided by fossil sources for example natural gas, coal and oil. Utilization of fossil fuels has caused an increase in CO_2 concentration. The resulting increase in CO_2 concentration in the earth's atmosphere is a cause of pollution and of climate change (Das & Veziro̅glu, 2001). The diminishing fossil fuel resources will eventually result in severe energy shortages in the future.

Problems involving climatic changes, together with increasing oil prices and the increase in government support to reduce pollution, are compelling and promoting the increase in renewable energy legislation, incentives and commercialization. Almost 85% of the world's energy requirement is provided by fossil sources, 7% from nuclear energy and 8% from renewable sources (Shell International, 2001).

It has been recommended by the Intergovernmental Panel on Climate Change (IPCC) that reducing global CO_2 emissions more than 50% be required to stabilise the CO_2 concentration in the atmosphere at 550 parts per million volumes (ppmv). In the Kyoto Agreements, which 106 countries have endorsed (Reith, Wijffels, and Barten, 2001) a number of developed nations committed to reduce their total CO_2 emission by 5% in 2010 relative to 1990 emission levels. Current evaluations show however that CO_2 emissions are constantly increasing and that such goals will not be met without more strategic and stringent measures

CO_2 emissions from the utilization of fossil fuels can be reduced by improved efficiency of electricity production and utilization. Alternatively, substituting fossil fuels (e.g. coal), which have a high CO_2 emission per unit of energy, with other fossil fuels with lower emissions (e.g. natural gas), for generation of electricity may be effective. Measures like these receive wide attention however they have been insufficient to reduce CO_2 emissions significantly. A more drastic approach is the utilization of renewable energy produced from hydropower, sunlight, wind and biomass. An example of a renewable energy produced from biomass is hydrogen, which is also a promising solution to the exhaustion of fossil fuel resources and also to reduce environmental pollution. It has also been recommended as the energy of the future (Das & Veziro˘glu, 2001). Hydrogen combines with oxygen during combustion and forms water as a by-product, 120 Megajoules of energy per kilogram of hydrogen (33.3kWh/kg) is released (Bose, Malbrunot, Benard, & Viola, 2007). Hydrogen is also well known as a non-polluting fuel because water, which is non-toxic, is the sole product of combustion (Suzuki, 1982).

Production of hydrogen by biological means is usually carried out at room temperature and a pressure of 1 atmosphere, and so the process is less energy demanding as compared with electrolytic and thermochemical processes for hydrogen production. Hydrogen can be produced by organisms which produce food from inorganic substrates (autotrophs) and by organisms, which depend on organic substrates for food nutrition (heterotrophs) (Das & Veziro˘glu, 2001).

There are different carbon sources that can be utilised by bacteria to produce hydrogen. Glucose can be utilised by different bacteria to produce hydrogen, for example *C. butyricum* (Masset et al., 2010) and *C. beijerinckii* (Chun-Mei et al., 2008). Whilst *Rhodopseudomonas sp.* has the ability to utilise starch, whey and dairy waste. *Rhodobacter sphaeroides* also has the ability to use lactic acid to yield H_2, O_2, CO_2 and small quantities of fatty acids (Das & Veziro⁻glu, 2001). *Rhodopseudomonas faecalis* RLD-53 utilised acetic acid and butyric acid, which are metabolic waste products of *Clostridium butyricum,* in the same fermentation vessel to produce hydrogen (Liu et al., 2010). Bio-hydrogen production by *Clostridium thermocellum* 27405, utilizing cellulosic biomass substrates have also been achieved (Levin et al., 2006). However these carbon sources have problems related to shortages, as the continuous use of some of the carbon sources will eventually lead to food shortages.

One possible source of biomass waste to resolve this problem that can also be used as a raw material for H_2 production is marine waste such as crab, shrimp and lobster shells which are made up of chitin. Chitin is composed of a long-chain polymer of N-acetlyglucosamine (GlcNAc) (Jolles & Muzzarelli, 1999), which is the second most abundant polymer in the world after cellulose (Watson, Zinyowera & Moss, 1998). Chitin occurs in nature in combination with other polysaccharides and proteins. Biological degradation of chitin is achieved with the use of endo- and exo- enzymes known as chitinases and β-N–acetylohexaminidases (Donderski & Swiontek, 2003). The usual way of disposing of this waste material is by landfill, incineration or release into the ocean. Due to increasing demands for energy, chitinous wastes represent a potentially valuable resource.

Poulicek and Jeuniaux, (1991) reported that an estimated value of 9.07×10^{13} Kg of chitin is produced annually and a substantial amount of it is produced in the oceans. Methods to utilize chitin or its monomer, N-acetylglucosamine, to produce H_2 have not been extensively researched on (Evvyernie et al., 2001) as well as optimizing the process for utilisation of GlcNAc for H_2 production.

A significant transition to dependency on renewable energy is not likely in the short term, because renewable energy production, operating with the current technology is not competitive with fossil established energy production. However, the production costs for renewable energy is on a downward trend as a result of technological improvements (Reith, Wijffels, and Barten, 2001).

The future energy economy will have an important role for hydrogen energy source. It can be used in fuel cell devices used to power vehicles and for distributed or decentralized electricity generation used in stationary fuel cell structures. Hydrogen can be converted to electricity in fuel cell devices very efficiently, thereby giving only water as a by-product, consequently reducing CO_2 and other fossil fuel pollutants. An important characteristic of the fuel cell technology is that very efficient electricity generation is achievable compared to other technologies. Hydrogen could potentially become a key energy source in 10 to 20 years, at the initial stages in the transportation sector using fuel cell devices and later on for distributed or decentralized generation of electricity

(Gosselink, 2002).

Extensive implementation requires the improvement of cost effective fuel cell technology expertise, hydrogen storage facilities and related structures.

Firstly, hydrogen will mainly be produced through the process of electrolysis or from fossil fuels sources in small- scale reformers as well as in large-scale, centralized plants for example through the process of steam-reforming of natural gas ($CH_4 + 2H_2O \rightarrow 4H_2 + CO_2$). Production on a large scale will permit recovery of the CO_2 to be used in greenhouses to encourage plant growth or for storage in reservoirs underground or stored in chemical form for example as carbonates. Hydrogen production derived from fossil fuels alongside CO_2 capture is important in the clean green fossil approach, intended at reducing the negative climate effects of the use of fossil fuels. This method will be an important part in the transitional period to a total renewable energy economy. Energy saving measures will be improved and the growth of highly efficient conversion technology like fuel cells for the use of natural gas and other fossil fuels may have a vital role in future (Shell International, 2000).

Though hydrogen produced from fossil sources may cause a significant reduction of emissions, the energy efficiency from the stage of production-to-end-use (natural gas → hydrogen → electricity) is limited by the loss of energy during the hydrogen production phase with CO_2 capture. Direct hydrogen production is therefore preferable from renewable sources such as the electrolysis of water by renewable electricity, or through the use of biomass gasification or biological hydrogen production. Through large-scale operation of renewable energy production, hydrogen can become a clean energy carrier for storage and transport. Electricity produced from renewable sources, for example hydropower, can be converted to hydrogen through the process of electrolysis aimed at storage and transport in liquid or gaseous form to consumers, to then be re-converted to electricity using fuel cells. As a result of this characteristic, hydrogen and electricity are likely to be the main clean energy carriers (EU European commission community research, 2002).

1.1.2 Role of biotechnology and biomass for green energy production

Some renewable energy sources are obtainable in the form of chemical energy stored in biomass. This biomass plays a key role as a feedstock for green production of gaseous and liquid biofuels. Wet or dry biomass can be used for energy production depending on the process involved for conversion into energy. An example of a dry biomass is wood and wet biomass includes the organic fraction of domestic waste, wastewater, agro-industrial wastes and slurries. Dry biomass is used for thermal conversion processes for example green electricity generation through the process of combustion or gasification. Dry biomass can also be used for the production of renewable diesel fuel by gasification, followed by Fischer-Tropsch synthesis, which helps to convert gasses for example a mixture of carbon monoxide and hydrogen (Hamelinck et al., 2003).

Wet biomass and residues can not be used as raw materials for thermal conversion because the preparation methods involved for the process includes transport and drying of the raw material which are uneconomical and requires large quantity of energy, which leads to more CO_2 production that provides a negative climate effect. Wet biomass can be consumed, as feedstock for renewable energy production. Biotechnological energy production processes are useful because they involve microorganims such as *Clostridium beijerinckki, C. paraputrificum* and *C. butyricum* at low temperature and pressure. Additionally these practices are suitable for decentralized energy production in small-scale systems located close to the availability of biomass or waste, thereby avoiding energy expenditure and costs for transport. Biotechnological processes related to energy production will generally be involved in the production of renewable biogas (methane and hydrogen) and liquid biofuels, (bioethanol and Acetone-Butanol-Ethanol) (Claassen et al., 1999; Kosaric et al., 1995).

1.1.3 Biotechnological production of bio hydrogen

Hydrogen formation and consumption are disconnected during biological hydrogen production process; which permits hydrogen to be available as the final product. These biological hydrogen production processes are currently in the research and development phase and extensive research is required to encourage commercialization of the technology. Various processes are presently under development, for example biomass photo biological processes and dark fermentation processes through which hydrogen can be produced with or without sunlight.

8

Another area to be investigated is the development of cheap and efficient hydrogen catalysts centered on the catalytic site of the hydrogen production enzymes, for example hydrogenases (Reith, Wijffels and Barten, 2001). The catalytic sites contain inorganic elements; mostly nickel, sulphur and iron. This information offers an opportunity for the production of affordable, synthetic hydrogen catalysts, which will be substituted for the costly precious metal based catalysts currently used for the inter- conversion of hydrogen and electricity in fuel cells devices and electrolysers which break water down into hydrogen and oxygen gasses. An alternative technique is the use of purified hydrogenase enzymes as part of the components of the hydrogen electrodes. However, this new area of development may contribute significantly to the future hydrogen economy, through the clever manipulation of natural systems (Cammack, Frey & Robson, 2001). Bio-hydrogen production and anaerobic digestion fermentation processes, which produce CO_2-neutral biogas from the utilization of biomass and some metabolites, can thus be competing technological applications (Reith, 2001). For productive utilization of feedstock it is imperative to use the optimal method for further improvement and implementation. This is contingent on several considerations including the total energy efficiency of the process, integrating or combining it into the energy infrastructure, its environmental impact (especially to reduce climatic effects by curbing CO_2 effect), production costs. Bio-hydrogen production processes have some important advantages over fossil fuel generation, for example the ability for CO_2 capture in the production stages which is a great opportunity to reduce global CO_2 emissions.

The introduction and utilization of renewable energy and also the continuing development of a hydrogen economy need optimal use and continuous improvement and modification of existing structures. Introducing the use of bio-hydrogen can perhaps be achieved through a transition strategy using the technological knowledge in the area of bio-methane production and the present gas infrastructure.

1.2 Hydrogen Production by biological process

As reported by Das & Veziro̅glu (2001), the production of hydrogen by biological processes can be classified as below:

- Biophotolysis of water by cyanobacteria and algae.

- Decomposition of organic compounds by photosynthetic bacteria.

- Production of hydrogen from organic compounds by fermentative bacteria.

- A combined process using both photosynthetic and fermentative bacteria.

1.2.1 Biophotolysis of water by Cyanobacteria and algae

The process of biophotolysis, results in the dissociation of water into molecules of hydrogen and oxygen as a consequence of the effect of light on biological systems (Yu & Takahashi, 2007). Microalgae produce hydrogen through this process. Photosynthesis is related to microalgal biophotolysis of water, in that carbon-containing biomass is produced during photosynthesis, while in biophotolysis hydrogen is generated (Das & Veziroḡlu, 2001). Das & Veziroḡlu, 2001 noted that the photosynthetic process used by microalgae involves 2 photosystem stages:

- Photosystem I (PS I), which facilitates the generation of reductants, which are utilized, for CO_2 reduction.

- Photosystem II (PS II), which involves water splitting and O_2 evolution.

For the photosynthetic process in green plants, only the reduction of CO_2 occurs; but have photosystem II however hydrogenase enzymes that catalyse the production of hydrogen are absent.

The microalgae that undergo biophotolysis possess chlorophyll *a*, as well as other pigments used to capture and absorb light energy for both of the photosystems. Such microalgae use the photosystems to carry out photosynthesis in a similar fashion to oxygenic plant-like photosynthesis (Yu & Takahashi, 2007). Certain microalgae like the Eukaryotic (green) and Prokaryotic cyanobacteria produce hydrogen with the help of a hydrogenase enzyme under certain conditions (Das and Veziroḡlu, 2001).

Pigments in the PSII capture photons forming an oxidant able to split water into O_2, protons (H+) and electrons (e-). The electrons pass through the photosystem along a progression of electron carriers and a cytochrome complex to PS1. PS1 pigments also capture photons to increase the energy state of the electrons, which then reduce oxidised ferredoxin or nicotinamide adenine dinucleotide phosphate ($NADP^+$). A proton gradient that develops across the cellular membrane encourages ATP formation through the ATP synthase. ATP is produced and NADPH is used to reduce the CO_2 produced through either the reductive pentose phosphate pathway or the Calvin cycle that enables cell growth. Excess carbon (reduced form) is stored as carbohydrate or lipid. Ferredoxin serves as an electron carrier, and under certain conditions it can also be used by hydrogenase or nitrogenase to reduce protons for hydrogen production (Yu & Takahashi, 2007). In addition the rate of production of hydrogen will be lower than the rate of reduction of CO_2, this is because low concentrations of O_2 inhibit the activity of the enzyme hydrogenase, thereby reducing the evolution of hydrogen during biophotolysis (Benemann, Berenson, Kaplan & Kauren, 1973).

1.2.2 Decomposition of organic compounds by photosynthetic bacteria

During the culture of anaerobic bacteria, photoheterotrophic bacteria, such as *Rhodobacter sphaeroides*, can utilise and reduce organic substrates like organic acids, through photosynthetic decomposition, to produce hydrogen (Venkataramana, Anoop, Nagamany & Maung, 2008). Organic acids and H_2S act as electron donors (Ida, Marcel, Jorge, Renee & Wijffels, 2002). Organic acids are utilised as substrates by the purple non-sulphur bacteria (PNS), whilst green and purple sulphur bacteria use sulphur compounds such as H_2S (Venkataramana, *et al.*, 2008). It is known that photosynthetic bacteria use the reductant H_2 for fixation of CO_2 and are also able to fix molecular nitrogen. Nitrogenase is known to be involved in the nitrogen reduction as well as ATP-dependent H_2

production. It was proposed that the production of H_2 occurs when cells produce excess ATP and the reducing ability of the cells exceeded the demand because of the ready availability of a carbon source or a reduced nitrogen source, such as glutamate/ aspartate (Nandi & Sengupta, 1998). Electron carriers carry electrons released from the utilisation of substrates and a proton gradient is formed, during electron transport, when protons are transported across the membrane concomitantly. At this point the enzyme ATP synthase generates ATP. The ATP generated is then used to transport electrons to ferredoxin. During growth in presence of low concentrations of nitrogen, these electrons can be utilized with extra ATP by the enzyme nitrogenase to reduce nitrogen to ammonium. But when molecular nitrogen is absent the enzyme nitrogenase reduces protons to produce hydrogen gas with an extra ATP energy utilized from the electrons derived from ferredoxin. High concentrations of organic acids can be converted into hydrogen gas and CO_2 (Ida et al., 2002).

In general, the decomposition of organic compounds by photosynthetic bacteria has a high theoretical hydrogen conversion yield. According to Das & Veziro̅glu (2001) the biochemical pathways for the photosynthetic decomposition process can be shown as follows:

$$(CH_2O)_2 \rightarrow Ferredoxin \rightarrow Nitrogenase \rightarrow H_2 \dots\dots\dots (1.1)$$
$$\uparrow \qquad\quad \uparrow$$
$$ATP \qquad ATP$$

1.2.3 Production of hydrogen from organic compounds by fermentative bacteria

The production of hydrogen can occur anywhere in the environment during anaerobic conditions, that is when no oxygen is present as an electron acceptor. The production of H_2 by fermentative bacteria has great advantages over production of H_2 by photosynthetic bacteria as hydrogen can be produced during the day and night with no light energy requirement, whilst yielding high hydrogen production and cell growth rates. During heterotrophic metabolism, energy yielding oxidation reactions take place anaerobically, and so bacteria search for mechanisms to maintain electrical neutrality. In other words when bacteria grow heterotrophically, these organic substrates are broken down by the process of oxidation to provide metabolites, bacterial growth and energy, as well as for electrons (de Vrije & Claassen, 2003). Consequently, strict and facultative anaerobic bacteria have specific control systems, which normalize and balance the electron concentration during metabolism. A part of these control systems involves the ability of these organisms to remove or expel excess electrons in the form of hydrogen (H_2) with the help of hydrogenase enzymes (Das & Veziroˉglu, 2001).

Under aerobic conditions oxygen acts as the electron acceptor during oxidative phosphorylation and it is reduced to water as the final product, whilst under anaerobic conditions other compounds will act as electron acceptors. For example protons, which will be reduced to hydrogen (H_2) can be used as electron acceptors (Yu & Takahashi, 2007). Nitrate and sulphate are other examples of anaerobic electron acceptors and the final reduction products of these are nitrogen gas and dihydrogensulphide (H_2S) (de Vrije & Claassen, 2003). Metabolites in fermentation processes can also act as electron acceptors. The ability for other electron acceptors to be reduced other than oxygen, requires the presence of nitrate reducing bacteria which possess a complex set of enzymes catalyzing the reduction of nitrate to nitrogen or the presence of a specific enzyme system of the hydrogen producing bacteria which possess the hydrogenase enzyme. Theoretically when glucose is utilized a maximum of 4 moles of H_2 per mole of glucose can be produced, along with energy (206KJ per mole of glucose) for growth (de Vrije & Claassen, 2003).

$$C_6H_{12}O_6 + 4H_2O \rightarrow 2CH_3COO^- + 2HCO_3 + 4H^+ + 4H_2 \dots\dots\dots\dots(1.2)$$

Some of the remaining hydrogen from glucose is stored in the various metabolites that are also produced. Figure 1.1 illustrates the pathway for hydrogen production from glucose.

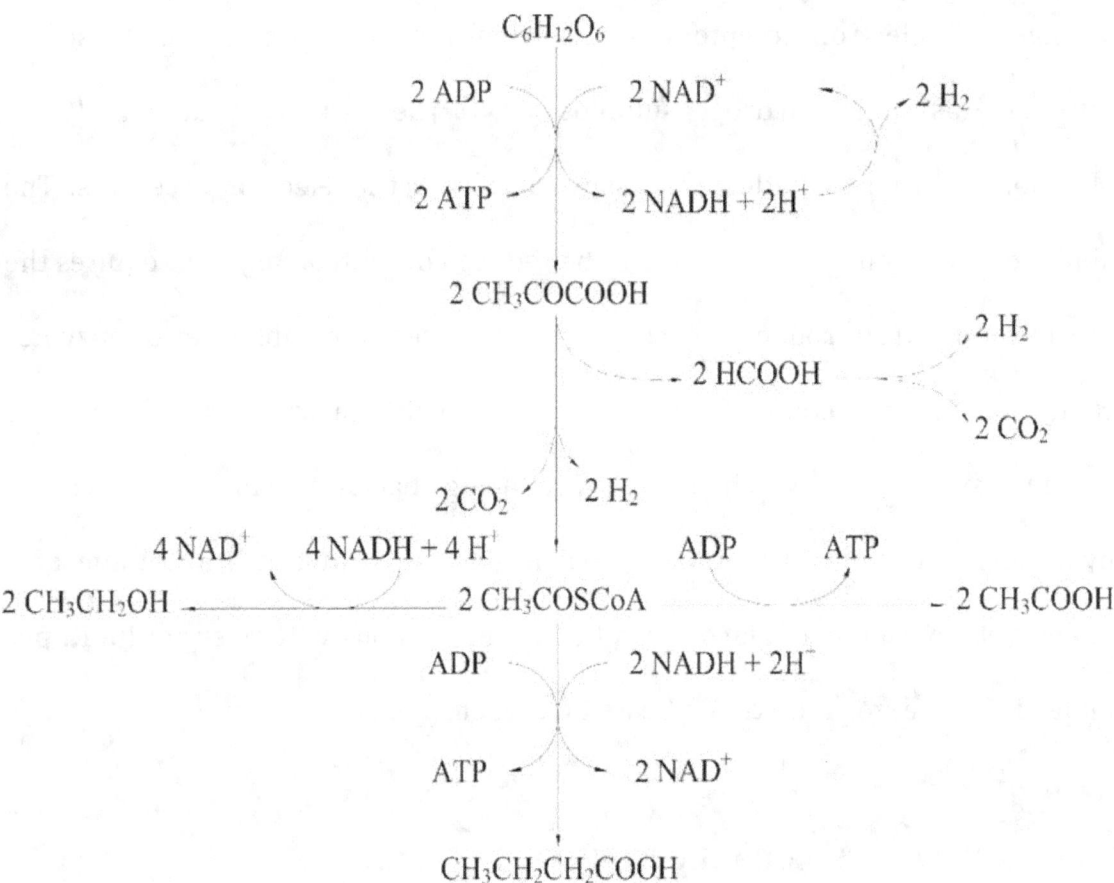

Figure 1.1 Hydrogen production pathways from the utilisation of glucose (Chenlin & Herbert, 2007).

Alongside H_2 some of the products of the fermentation process include organic acids such as acetate, butyrate, formate, ethanol and propionate. The stoichiometric relationships between the substrate, glucose, and the various metabolites are analyzed in the equations given below.

Acetic acid

$$C_6H_{12}O_6 + 2H_2O \rightarrow 2CH_3COOH + 2CO_2 + 4H_2 \ .... (1.3)$$

Butyric acid

$$C_6H_{12}O_6 \rightarrow CH_3CH_2CH_2COOH + 2CO_2 + 2H_2 \ ....(1.4)$$

Ethanol

$$C_6H_{12}O_6 + 2H_2O \rightarrow CH_3CH_2OH + CH_3COOH + 2H_2 + 2CO_2 \dots\dots(1.5)$$

Propionate

$$C_6H_{12}O_6 + 2H_2 \rightarrow 2CH_3CH_2COOH + 2H_2O \dots\dots\dots\dots\dots\dots(1.6)$$

The reactions shown in equations 1.3 and 1.4 yield 4 moles of hydrogen from each mole of glucose when acetate is produced, and 2 moles of hydrogen from each mole of glucose when butyrate is produced (Nandi & Sengupta, 1998), 2 moles of hydrogen per mole of glucose are also produced in the ethanologenic fermentation shown in equation 1.5 (Chenlin & Herbert, 2007). Under some environmental conditions hydrogen may subsequently be consumed in the production of organic acids (propionate) as shown in equation 1.6 (Chenlin & Herbert, 2007). As reported by Vavilin, Rytow, & Lokshina (1995), using propionic and butyric acid producing bacteria, it was observed that there was a decrease in hydrogen production, which occurred simultaneously with the initiation of propionic acid production.

Glucose can be converted to pyruvate through the glycolytic pathway, thereby producing adenosine triphosphate (ATP) from adenosine diphosphate (ADP) during the production of butyrate as shown in figure 1.1. With the enzymes pyruvate ferredoxin oxidoreductase and hydrogenase, pyruvate is converted to acetylcoenzyme A (acetyl-CoA), carbon dioxide CO_2 and hydrogen. Under different culture conditions, pyruvate can also be converted to acetyl-CoA and formate (Chenlin & Herbert, 2007). This is shown in the equation below:

$$CH_3COCOOH + H\text{-}CoA \rightarrow CH_3 CO\text{-}CoA + HCOOH \quad\text{...(1.7)}$$

Das and Veziro̅glu (2001) reported that hydrogen can also be produced through formate decomposition under some fermentation conditions, for example during mixed acid fermentation or butane 2, 3 diol fermentation.

$$HCOOH \rightarrow H_2 + CO_2\text{..(1.8)}$$

Under certain environmental conditions acetyl-CoA may be converted to acetate, butyrate and ethanol. For example:

Acetic acid

$$CH_3CO\text{-}CoA + H_2O \rightarrow CH_3COOH + H\text{-}CoA; \text{.. (1.9)}$$

Ethanol

$$CH_3CO\text{-}CoA + 2NADH + 2H^+ \rightarrow CH_3CH_2OH + H\text{-}CoA + 2NAD^+\text{.............................. (1.10)}$$

(Das & Veziro̅glu, 2001).

The NADH produced during the conversion of glucose to pyruvate may also be used in the production of butyrate and acetate with the generation of ATP from acetyl-CoA (equation 1.11).

$$C_6H_{12}O_6 + 2NAD^+ \rightarrow 2CH_3COCOOH + 2NADH + 2H^+\text{.......................................(1.11)}$$

Acetate and butyrate are the most common metabolites derived from the fermentation of carbohydrate (Chenlin and Herbert, 2007). In addition, hydrogen can also be produced with the help of NADH through the re-oxidation of NADH in the NADH pathway (Das and Veziro¯glu, 2001). In other words residual NADH is oxidized thereby evolving hydrogen and NAD. For example:

$$NADH + H^+ \rightarrow H_2 + NAD^+ \dots\dots\dots\dots\dots\dots\dots\dots\dots\dots\dots\dots\dots(1.12)$$

The hydrogenase enzymes that contain Fe catalyse hydrogen production while those that contain Ni and Se enable the uptake of hydrogen. Hydrogen production during fermentation can be affected or influenced by the partial pressures of the products expressed (Chenlin and Herbert, 2007).

1.2.4 Combination Process using the photosynthetic and fermentative bacteria

It may be possible to combine sugar fermentation processes of fermentative bacteria, that do not require light energy, and the organic acid metabolism of photosynthetic bacteria, which use light as a source of energy. This combination or hybrid process improves the yield of hydrogen from a carbon source. Fermentative bacteria produce hydrogen and organic acids during the process of fermentation and those organic acids produced during fermentation can go on to serve as a carbon source for photosynthetic bacteria to produce hydrogen. Photosynthetic bacteria use light energy to degrade organic acids and enable hydrogen production, and consequently the combination process increases hydrogen production.

During biophotolysis minute quantities of O_2 can inhibit hydrogenase activities, which decrease hydrogen production. The difficulty in achieving continued H_2 photo-production is that the photosynthetically produced O_2 irreversibly inactivates the production of H_2 (Yu & Takahashi, 2007). Although the decomposition of organic compounds using photosynthetic bacteria lacks the system of O_2 production, which causes the problem of O_2 inactivation, the photofermentation process does not have the ability to continue fermentation during the absence of light.

Fermentative hydrogen production from organic compounds has a high rate of hydrogen production, because it is capable of producing H_2 from organic substrates constantly in light and dark conditions. Therefore fermentative hydrogen production would seem to be an ideal candidate for industrial hydrogen production (Das and Veziro˘glu, 2001).

1.3 Influence of Acetate and Butyrate yields on biohydrogen production

The pH of a culture medium is an important factor influencing most organic acid fermentations, such as butyric acid (Vandak et al., 1997), propionic acid (Hsu and Yang, 1991), lactic acid (Silva and Yang, 1995), and acetic acid (Tang et al., 1989). Cell growth and fermentation rate are influenced by culture medium pH as well as affecting changes in final metabolic product yield. Butyric acid fermentation, by *C. tyrobutyricum* utilising xylose changed from a major butyrate production at medium culture pH of 6 to predominantly lactate and acetate production at medium culture pH of 5 (Zhu & Yang, 2004). Nandi and Sengupta (1998) reported that 4 moles of hydrogen are produced from each mole of glucose when acetate is produced, and 2 moles of hydrogen from each mole of glucose when butyrate is produced (Nandi & Sengupta, 1998). In experiments carried out by Fang and Liu (2002), increase of culture medium pH from 4.0 to 7.0 in a mixed culture resulted in the decrease of butyrate but in the increase of acetate production when glucose was utilized.

Hydrogen production is normally followed by organic acid production as well as solvent production. The production of these metabolic products suggests microbial metabolic pathway shifts.

Reports by Khanal et al. (2004) showed that a gradual production of organic acids resulted in a decline in medium pH to about 5.5, and this occurred before the production of hydrogen was initiated. The acetate to butyrate ratio during the production of hydrogen showed a similar pattern. The maximum acetate to butyrate ratio was detected during the exponential growth phase. When the stationary phase of hydrogen production was reached the acetate to butyrate ratio declined. At lower initial pH, the maximum acetate to butyrate ratio was higher.

Lay (2000) reported that microbial shift from hydrogen or acid production to solvent production usually starts at culture medium pH levels of 4.1 or lower. Above this pH range the metabolic pathway changes, which is usually the cause

21

for hydrogen production decline and not a reduction in hydrogenase activity.

These changes in organic metabolite ratios suggest a metabolic change due to pH changes, partial pressure of hydrogen (hydrogen content in the biogas) and the accumulation of metabolic products. (Khanal et al. 2004).

1.4 Hydrogen producing micro- organisms

1.4.1 Strict anaerobes

1.4.1.1 Fermentative bacteria: clostridia

A variety of features have been used to classify clostridia, which include: the capability of the bacteria to form endospores; the capability to carry out only anaerobic energy metabolism; and the possession of a Gram-positive cell wall. In addition the cells of all clostridia species are rod-shaped with round or pointed ends (Minton & Clarke, 1989). The genus *Clostridium* comprises obligate anaerobes that generate energy in the form of ATP through substrate-level phosphorylation during the process of fermentation, as clostridia lack the cytochrome system for oxidative phosphorylation (Nandi & Sengupta, 1998). During the process of glycolysis glucose is broken down to pyruvate with the generation of ATP and NADH. Pyruvate in turn is metabolized to produce acetyl-CoA, CO_2 and H_2 with the help of the enzyme pyruvate ferredoxin-oxidoreductase and hydrogenase (Chenlin & Herbert, 2007). The NADH produced can also be used in the formation of butyrate from acetyl CoA with the generation of ATP by the enzyme phosphobutyrylase and butyrate kinase (Thauer, Jungermann &

Decker, 1977). Depending on the particular fermentation conditions and also the bacterial species, NADH may also be used to produce ethanol from acetyl-CoA (Das & Veziroglu, 2001). Energy in the form of ATP can also be generated from acetyl-CoA by acetyl kinase whilst NADH is oxidised by ferredoxin-oxidoreductase, ferredoxin and hydrogenase. For example, in the hydrogen production pathway, ATP is produced from the conversion of acetyl-CoA to acetate (Chenlin & Herbert, 2007).

A large number of anaerobic bacteria produce hydrogen from the utilization of hexoses during acetic and butyric acid fermentations as well as acetone-butanol-ethanol fermentations. Mixtures of metabolic organic products are produced by Clostridia and H_2 metabolized from glucose is determined by the butyrate/acetate ratio.

The strictly anaerobic bacterium *C. beijerincki* AM21B isolated from termites yielded 1.8 to 2.0 moles H_2 from metabolizing glucose as its main substrate (Taguchi, Chang, Takiguchi & Morimoto, 1992). *C. beijerincki* AM21B strain has the ability to utilize various other types of carbohydrates, for example cellobiose, sucrose, xylose, arabinose, galactose and fructose with efficiencies from 15.7 to 19.0 mmol/g of substrate in batch fermentations processes with duration of 24 h (Taguchi, Chang, Mizukami, Saito-Taki, Hasegawa & Morimoto, 1993). Using starch as its main substrate H_2 was produced with equal efficiencies, but continuous production of H_2 was unattained. Hydrogen production stopped before the exhaustion of the carbon sources in the medium. Hydrogen was produced more efficiently from a different *Clostridium sp* strain N02 which was also isolated from termites. This *Clostridium sp* N02 utilised xylose and arabinose as its main carbon source to produce H_2 (13.7 and14.6 mmol/g or 2.1 and 2.2 mol/mol) more efficiently than using glucose as its main carbon source (11.1 mmol/g or 2.0 mol/mol) (Taguchi, Mizukami, Hasegawa & Saito-Taki, (1994). The ability of these Clostridia to produce H_2 from a variety of substrates indicates that both clostridium species can be used for the production of H_2 from utilizing cellulose and hemicellulose contained in plant biomass as its main source of carbon. The hydrolysis of biomass, which helps in breaking down chemical bonds and for easy utilization of biomass which leads to the production of utilizable, fermentable substrate can also be performed during fermentation, in a simultaneous saccharification and fermentation process. This preparation can also be done in a separate process prior to fermentation. In fermentative hydrogen production processes, it was observed that the simultaneous saccharification of xylan alongside a crude xylanase by *Clostridium* strain no. 2 could progress in a single fermenter (Taguchi, Mizukami, Yamada, Hasegawa & Saito-Taki, 1995). Though, in this process the simultaneous conversion was not as effective as compared to the separate conversion of the hydrolysate. Additionally, the cellulose hydrolysate (Avicel) with readily available cellulase preparation used in bio-hydrogen fermentation by strain no 2, could not be achieved in one flask because there was a significant difference in fermentation conditions necessary for growth and enzyme activity of strain no. 2. An alternative method is to utilize clostridia strains, which are capable of producing cellulase or xylanase activity. Presently, there are no strains available that are capable of hydrolyzing both glucans and xylans. In the report by Taguchi, Hasegawa, Saito-Taki and Hara (1996) a novel strain, *Clostridium* sp. strain X53 was isolated from wild termites, and when cultured in a batch process production of xylanase and conversion of xylan to produce hydrogen was observed. In contrast to xylose, the chemical reactions required for hydrogen production from xylan did not differ significantly, although the total yield derived from xylan was lower than the yield produced from xylose. Continuous hydrogen fermentation has been achieved by Clostridia utilizing glucose as the main substrate (Heyndrickx, Vos & Ley, 1990). It was reported that in both batch and continuous fermentation processes, the *Clostridium* sp. strain no. 2 had a maximal H_2 production rate and molar yield that were comparable. It was reported that in an aqueous two-phase system (polyethylene glycol and dextran) continuous hydrogen production was achieved during fermentation when there was a continuous production of hydrolysate. There was a higher H_2 production rate and the H_2 yield with Avicel hydrolysate compared to glucose. While utilizing Avicel the H_2 yield was higher, compared to the theoretical maximum, the combination of hydrolyzed dextran with the Avicel hydrolysate was implied (Taguchi, Yamada, Hasegawa, Taki-Saito & Hara, 1996). H_2 production rates from 20.4 to 21.7 mmol/L.h during continuous fermentation processes have been recorded (Heyndrickx, Vansteenbeeck, Vos & Ley, 1986) with low yields of 1.4 mole H_2 from 1 mole glucose. In other fermentation cultures higher yields of 2.4 mol/mol were achieved with lower H_2 production rates of 7 mmol/L.h (Taguchi, Mizukami, Saito-Taki, Hasegawa, 1995). Utilizing xylose as the main carbon source, comparable results were obtained with maximum H_2 production rates of

25

21.0 mmol/L.h were recorded, with yields of 1.7 H_2 mol/mol xylose. Waste products like chitin and its amino sugar N-acetylglucosamine have also been used to produce hydrogen. Hydrogen yields of 2.2 mol/mol N-acetyl-D-glucosamine have been achieved using chitinolytic bacterium *Clostridium paraputrificum* at production rates of 31 mmol/L.h (Evvyernie et al., 2000). The ability of Clostridia to utilize various substrates and produce H_2 appears promising. Continuous improvements in optimization process conditions will eventually increase hydrogen yields and production rates.

1.4.1.2 Rumen bacteria

Rumen bacteria are strictly anaerobic bacteria known to produce hydrogen. *Ruminococcus albus* is an example of a rumen bacterium known to produce H_2 as well as other organic acids like ethanol, acetate, formate and CO_2 from utilizing carbohydrates. Innotti, Kafkawitz, Wolin and Bryant (1973) reported that a H_2 yield of 2.4 mol/mol glucose was achieved in a continuous culture.

1.4.1.3 Thermophiles

An example of a thermophile is *Pyrococcus furiosus*, which is an Archaea. This Archaea produces H_2, organic acids and CO_2 from utilizing carbohydrates (Godfroy, Raven & Sharp, 2000). Extreme and hyper thermophiles, which utilize substrates and produce metabolized products, are also known to produce hydrogen from carbohydrates. Examples of cellulolytic thermophiles, extreme and hyper-thermophilic bacteria that produce hydrogen include species of

Dictyoglomus, Fervidobacterium, Spirocheta, Thermotoga, Anaerocellum, Caldicellulosiruptor, Clostridium, and *Thermoanaerobacter*. Batch fermentation cultures carried out by Schröder, Selig and Schönheit (1994) at 80°C with *Thermotoga maritima* recorded a hydrogen yield on glucose utilization of 4 mol/mol which was equal to the theoretical maximum value. Though there was a low cell density recorded (1.4 x 10^8 per mL) and low glucose consumption (1.6 mM), with a maximum hydrogen production rate of 10 mmol/h achieved. A similar stoichiometric result was achieved when results obtained by *T. maritima* were compared to two moderate thermophiles, *Acetothermus paucivorans* and *Acetomicrobium flavidum*, which were grown at 60°C (Reith, Wijffels and Barten, 2001). Other meaningful results during sugar fermentation processes by other extreme thermophiles have been published (van Niel, Budde, de Haas, van der Wal, Claassen & Stams 2002). During fermentation cultures of *Caldicellulosiruptor saccharolyticus* utilizing sucrose as its main carbon source at 70°C and *Thermotoga elfii* utilizing glucose at 65°C, stoichiometric yields of 3.3 mole H_2 per mole hexose were recorded. Maximum hydrogen production rates of 8.4 and 2.7 mmol/l/h, respectively, were also recorded. These results suggest that higher hydrogen yields produced from the utilization of hexose can be achieved by the fermentation activities of extreme and hyper-themophiles in comparison to mesophilic facultative and strict anaerobes.

1.4.1.4 Methanogens

Methanogens are microorganisms known to produce methane as a byproduct during metabolism in low oxygen environmental conditions. These methanogens are usually characterized by the involvement of the hydrogenase enzyme usually involved in the oxidation of H_2 linked to CH_4 production and CO_2 reduction (de Vrije & Claassen, 2003).

1.4.2 Facultative anaerobes

These are bacteria, which are able to produce energy by aerobic respiration and then switch back to anaerobic respiration depending on the concentration of oxygen and utilizable fermentable nutrients in the environment. Facultative anaerobes are usually unaffected by the presence of oxygen. These bacteria have the ability and advantage of rapidly consuming oxygen thus reestablishing anaerobic conditions in their immediate environment. Strict anaerobes are very sensitive to oxygen and often do not survive in the presence of low environmental oxygen concentrations.

1.4.2.1 Enterobacter

Enterobacter posses some valuable and beneficial characteristics that favor H_2 production. Other members of the family Enterobacteriaceae also posses similar qualities. These qualities include high growth rates and utilization of a wide variety of carbon sources. High pressures of produced hydrogen do not inhibit hydrogen production by Enterobacter (Tanisho, Suzuki & Wakao, 1987). Nevertheless, the H_2 yield derived from the utilization of glucose is usually lower in comparison to that of Clostridia. In a report by Tanisho, Wakao & Kosako (1983). *E. aerogenes* E.82005 was isolated from *Mirabilis jalapa* leaves. This strain was cultured under batch fermentation process, and a hydrogen production rate of 21 mmol/L.h was achieved over a period of 23 h with a H_2 yield of 1.0 mol/mol glucose. During a continuous fermentation process hydrogen was produced for a total duration of 42 days, using molasses as its main substrate using the same strain of *E. aerogenes*. An average H_2 production rate of approximately 17 mmol/L.h was achieved. Furthermore the average H_2 yield derived from the utilization of sucrose was 1.5 moles. In comparison to products produced in batch fermentations, lactate was identified as the major product while butyrate and acetate were produced in lower quantities (Tanisho and Ishiwata (1994). Production of hydrogen is not inhibited by high pressures, however flushing the Enterobacter fermentation culture with argon improved the H_2 yield to 1.6 mole per mole of glucose. It was concluded that the reason for the yield improvement was as a result of the removal of CO_2 after flushing the fermenter with argon (Tanisho, Kuromoto & Kadokura, 1998). In a report by Yokoi, Ohkawara, Hirose, Hayashi and Takasaki (1995), an aciduric *E. aerogenes*, strain HO-39 was isolated, that was capable of growing and producing hydrogen at low pH level of 4.5. In a different fermentation culture hydrogen was produced at an approximate rate of 5mmol/L.h for 26 days in a continuous culture. This was also achieved without pH control of the fermentation system. To improve on the H_2 production rates *E. aerogenes* and *E. cloacae* were genetically modified and the resulting mutants of *E. aerogenes* and *E. cloacae* were grown. These mutants were modified by preventing or blocking the characteristic of producing other metabolites, organic acid and alcohols. This was done because the general characteristic of *E. aerogenes* and *E. cloacae* to produce other metabolites reduces the production of hydrogen. Rachman et al. (1997) cultured an *E.*

aerogenes mutant that produced lower quantities of ethanol and butanediol, although the quantities of organic acids produced were similar to the wild type but the hydrogen yield and production were double the quantity produced by the wild type (Rachman et al., 1997). In an experiment to enhance the hydrogen production by *Enterobacter cloacae,* Kumar and Das (2000) isolated *E. cloacae* IIT- BT 08 strain from a leaf extracts. *E. cloacae* was able to grow and produce hydrogen by utilizing a variety of carbon sources. A maximum yield of 2.2 mol/mol of hydrogen was derived using glucose as a substrate in a batch fermentation process while a maximum of 6.0 mol/mol of sucrose was achieved and 5.4 mol/mol was achieved from utilizing cellobiose. Using sucrose as the main source of carbon a maximum H_2 production rate of 35 mmol/L.h was achieved. A similar method was used to develop a modified mutant of *E. aerogenes* to improve on hydrogen production (Kumar, Ghosh & Das, 2001). Batch fermentation cultures with these mutants were carried out and they produced less ethanol and butanediol as well as low yields of lactate and butyrate. In this experiment the resulting yield of acetate was comparable and similar to the wild type strain. The inhibition in the metabolic pathway for formation pathways of alcohols and organic acids resulted in a 1.5 times increase in H_2 yield on glucose utilization which was 3.4 mole per mole glucose. Continuous fermentations were carried out with *E. aerogenes* wild types and mutant (Rachman, Nakashimada, Kakizono & Nishio, 1998). A maximum production rate of 58 mmol/L.h was achieved at a dilution rate of 0.67 /h using the mutant in the fermentation process. This was almost double the production rate of the wild type. The molar hydrogen yield was stable at 1.1 when glucose was utilized as a substrate. In a report studying the process development of continuous hydrogen production by *Enterobacter aerogenes* in a packed column reactor *E. aerogenes* wild type strain produced hydrogen on a starch hydrolysate. The maximum yield was 1.5 mol/mol of H_2 from glucose in a continuous fermentation at a dilution rate of 0.1 h^{-1} (Palazzi, Fabiano & Perego, 2000).

1.4.2.2 *Escherichia coli.*

It is known that *E. coli* has the ability to utilize formate in the absence of oxygen to produce H_2 and CO_2. Enzymes involved in this process include formate dehydrogenase and hydrogenase (Gray & Gest, 1965). Continued lysis of formate needs the action of blocking of other anaerobic reductases (Nandi & Sengupta, 1996). Production of hydrogen from carbohydrates has been achieved during fermentation using *E. coli*. There are irregularities involved on the pathway leading to the production of hydrogen either through formate or without formate acting as an intermediate in the production process. The molar H_2 yield when glucose is utilized by culturing *E. coli* was 0.9 (Blackwood, Neish & Ledingham, 1956) or 1.2 by immobilized cells.

1.4.2.3 Citrobacter

Under anaerobic condition a *Citrobacter* species, *Citrobacter* sp. Y19 which was isolated from sludge digesters, has been known to produce hydrogen derived from CO and H_2O using the water-gas shift reaction (Jung, Kim, Park & Park, 2002). Production of hydrogen was monitored under batch conditions in serum bottle and also during continuous processes. In the continuous system hydrogen production rates of 15 mmol/l/h were recorded.

1.4.3 Aerobes

1.4.3.1 Alcaligenes

Kuhn, Steinbuchel and Schlegel, (1984), reported hydrogen evolution by strictly aerobic hydrogen bacteria under anaerobic conditions. They reported that strains and mutants involved in the fermentation process of the strictly aerobic hydrogen-oxidizing bacterium *Alcaligenes eutrophus* were cultured heterotrophically utilizing fructose or gluconate as a substrate and then exposed to anaerobic conditions in the presence of the organic substrates, molecular hydrogen was produced. Pinchukova, Varfolomeev and Kondrateva, (1979) reported that a soluble hydrogenase was isolated for *A. eutrophus*.

1.4.3.2 Bacillus

Kalia, Jain, Kumar and Joshi (1994) studied the fermentation of bio waste to H_2 by *Bacillus licheniformis*. The bio-waste contained wheat grains and other organic solids; these were then utilized by fermentation to produce H_2 during a batch fermentation process. This system produced 225, 205 and 203 l of biogas, a mixture of H_2, CO_2 and H_2S, per kg of organic solids respectively. Kumar, Jain, Sharma, Joshi and Kalia, (1995) also studied viable cells of the hydrogen producer *Bacillus licheniformis* and a mixed microbial culture, which were immobilized on brick dust and in beads of calcium alginate. In batch fermentation culture, the mixed culture cells yielded 8.2 l H_2 (0.36 mol H_2/mol glucose) of glucose utilized, however *B. licheniformis* produced 13.1 l H_2.

1.4.4 Co and mixed cultures

As reported by Yokoi, Tokushige, Hirose, Hayashi and Takasaki (1998), the combined potentials of *Clostridium butyricum* and *Enterobacter aerogenes* in a co-culture were investigated in a continuous fermentation system. The high hydrogen production yield of *C. butyricum* and the characteristic of oxygen consumption by the facultative anaerobe *E. aerogenes* were combined. The activity of *E. aerogenes* was enough to quickly reestablish anaerobic environmental conditions in the fermenter with short oxygen exposures. During a continuous fermentation process-utilizing starch on porous glass beads with immobilized mixed cells exhibited a H_2 production rate of almost 50 mmol/L.h and a final H_2 yield of 2.6 utilizing glucose at dilution rates of 1 h^{-1}. There are various sources to isolate microflora for mixed cultures. These sources include sludge from anaerobic digesters of municipal sewage or organic waste and sludge from kitchen wastewater etc. These microflora sometimes include unwanted bacteria for example methanogens, which utilize the metabolized hydrogen to produce methane. In the preparation of enrichment cultures for the microflora, these cultures are usually treated by forced aeration of the sludge or by heat treatment, which helps to inhibit the action of the hydrogen consumers allowing the spore forming anaerobic bacteria to grow and survive. Using mixed cultures industrially for the production of hydrogen from organic waste can be more advantageous over the use of pure cultures, as it can be easy for pure cultures to become contaminated with the involvement of the hydrogen-consuming bacteria. Sparling, Risbey & Poggi-Varaldo, (1997) investigated hydrogen production from inhibited anaerobic composters and reported that hydrogen production after wastewater treatment inhibited methane production followed by low H_2 yields and lack of consistency (Sparling et al., 1997). They stated that acetylene was effective in inhibiting the activities of the methanogens involved in the batch anaerobic system inoculated with an undefined cellulotytic consortium isolated from an anaerobic digester. The rate and quantity of hydrogen produced were not affected by acetylene in pure cultures of *Clostridium thermocelium* cultured under the same conditions. Ueno, Kawai, Sato, Otsuka & Morimoto (1995) reported that anaerobic microflora from sludge compost utilized cellulose to produce hydrogen with a high yield of 2.4 mol/mol hexose during a batch fermentation experiment cultured at 60°C. Sewage sludge was used to utilize glucose and sucrose to produce hydrogen in a continuous system at 35°C (Chen, Lin & Chang, 2001). Molar H_2 yields of 1.7 and 3.4 were achieved on the utilization of glucose and sucrose, respectively. A production rate of approximately 26-29 mmol/L.h was recorded over 14 days. A kinetic model was used to explain and to also predict the results and based on the results achieved, it was assumed that product formation was fundamental and was largely a linear function related to biomass concentration. Lay (2000) worked with mixed culture isolated from digester sludge. Culturing conditions were varied using a statistical optimization design, which is the central composite design, this was done, so as to model and optimize the anaerobic digested sludge, thereby converting starch and cellulose to produce hydrogen. Mizuno, Dinsdale, Hawkes, Hawkes and Noike (2000) enhanced the H_2

production yield from mixed cultures isolated from fermented soybean meal by sparging the culture medium continuously in the reactor with oxygen free N_2 gas when glucose was utilized. Throughout the 8-week duration of continuous operation steady H_2 production rates of approximately 8 mmol/L.h were achieved. Noike and Mizuno (2000) also reported their work with the use of organic waste as substrate, using the same mixed culture in a batch fermentation system. They achieved a hydrogen production yield of 1.7 to 2.5 mol/mol from utilizing hexose. Carbohydrates were utilized more while the proteins were not degraded.

Comparing the hydrogen yield produced by the strictly anaerobic clostridia to the facultative anaerobic clostridia, results showed that the strictly anaerobic clostridia produce a higher hydrogen yield of approximately 2 mol/mol, than facultative clostridia, which produce hydrogen yields generally less than 2 mol/mol. Higher hydrogen molar yields of greater than 3 have been achieved with the use of genetically modified mutants of Enterobacter (de Vrije & Claassen, 2003). These mutants were blocked at their biosynthetic pathways, which are involved in the production of organic acid and alcohol. In fermentations performed with mixed cultures, molar H_2 production yields of about 2 are achieved. This reveals the dominant presence of clostridia involved in the fermentation process when in enriched cultures. However the production and formation of product seems to be dependent on bacterial cell density. Furthermore thermophiles generally grow to low cell densities; consequently hydrogen production rates are projected to be at low quantities. A maximum of

23 and 58 mmol/l have been reported for Clostridia and Enterobacter respectively. Though the rates of hydrogen production by co and mixed cultures achieved are approximately 30-50 mmol/L.h (de Vrije & Claassen, 2003).

1.4.5 *Clostridium beijerenckii*

The bacterium *C. beijerinckii* is a strictly anaerobic, motile, rod shaped bacterium that is mesophilic. It produces certain metabolites during the process of fermentation; these metabolites include carbon dioxide, hydrogen gas, acetate, lactate, butyrate, butanol, ethanol, acetone, acetoin and methyl carbonile. *C. beijerinckii* exhibits a long lag phase, with a high initial concentration of glucose required to progress to an exponential phase during fermentation (Tae-Young, Gi-Cheol, Sung and Suk, 2008). *C. beijerinckii* AM21B strain isolated from termites was used for hydrogen production from utilization of starch during fermentation. It was recorded that the strain produced a total of 2450 ml (109.37 mol) and 2255 ml (100.66 mol) of H_2 from glucose (10g) and starch (10g), respectively, with a maximum rate of evolution at 660 ml and 410 ml of H_2/h (29.46 and 18.30 mol of H_2/h), respectively. This fermentation process was carried out in 1 l of peptone yeast (PY) medium with a glucose or starch concentration of 10 g/l at pH 6.5 (Fumiaki, Jun, Shuya & Masayoshi, 1992). Simunek, Tishchenko & Koppova (2008) reported that *C. beijerinckii* has the ability to degrade colloidal chitin, and consequently it must express a range of chitinolytic enzymes. Indeed, endochitinase, exochotinase and chitosanase activities were detected in the cultivation medium. High intracellular activity of

N-acetyl-β-glucosaminidase (GlcNAc-ase) and chitosanase activities were also detected.

1.4.6 *Clostridium paraputrificum*

C. paraputrificum is a strictly anaerobic, mesophilic and chitonolytic bacterium. The organism can utilise starch and glucose as substrates as well as N-acetylglucosamine, a monosaccharide of chitin. It is rod shaped with a cell size of 1 by 5 to 10μm and is Gram positive, but turns negative rapidly with incubation. When utilising metabolic substrates such as glucose *C. paraputrificum* produces large volumes of gasses and some organic acids. During the fermentation of chitin it produces hydrogen and carbon dioxide in an approximate ratio of 3:1 along with some organic acids (Evvyernie, et al., 2000). Organic acids such as acetate (21.6 mmol/l) and lactate (18.9 mmol/l) have been reported to be produced in larger quantities than propionate (1.7 mmol/L) and butyrate (2.6 mmol/L) when chitin in colloidal suspension is used as a substrate (Simunek, Kopecny, Hodrova & Bartonova, 2002). When *C. paraputrificum* was cultivated in a 1 l fermenter under optimised growth conditions it doubled in biomass every 30 minutes when using N-acetyl glucosamine as a substrate. It also produces hydrogen (1.9 mol H_2/mol GlcNAc) at a pH of 6.5, a temperature 45°C, and an agitation rate of 250 rpm (Evvyernie, et al., 2000).

Using N-acetylglucosamine and ball-milled chitin as a substrate during cultivation, *C. paraputrificum* was found to produce acetate and propionate as the major organic acid products, and lactate and butyrate as the minor ones

(Evvyernie et al., 2000). Furthermore, it was reported that *C. paraputrificum*, strain M-21, elicited 2.2 and 1.5 mol of H_2 gas from 1 mole of N-acetylglucosamine and ball milled chitin (equivalent to 1 mole of GlcNAc), respectively, when cultivated at a pH of 6. The chitinases ChiA and ChiB were also detected during the fermentation process (Evvyernie, Kenji, Shuichi, Tetsuya, Kazuo & Kunio, 2001). Table 1.1 shows differents approaches and substrates used for biohydrogen production.

Table 1.1 Showing different microorganisms and carbon substrates used for bio-hydrogen production.

Organisms Used	Carbon Substrate	Moles of H_2/ Mole of substrate	References
Facultative bacteria: Clostridia			
Clostridium beijerinckii	Glucose	1.8 to 2.0	Taguchi et al. (1992)
Clostridium beijerinckii	Glucose	2.52	Chun-Mei et al. (2008)
Clostridium specie NO2	Xylose	2.1	Taguchi et al. (1994)
Clostridium specie NO2	Arabinose	2.2	Taguchi et al. (1994)
Clostridium specie NO2	Glucose	2.0	Taguchi et al. (1994)
Clostridium butyricum	Glucose	1.4	Heyndrickx et al. (1986)
Clostridium paraputrificum	Chitin	1.5	Evvyernie et al. (2001)
Clostridium paraputrificum	N-acetylglucosamine	2.2	Evvyernie et al. (2001)
Clostridium paraputrificum	N-acetylglucosamine	1.9	Evvyernie et al. (2000)
Rumen bacteria			
Ruminiciccus albus	Glucose	2.4	Innotti et al. (1973)
Thermophiles			
Thermotoga maritima	Glucose	4.0	van Niel et al. (2002)
Thermotoga elfii	Hexose	3.3	van Niel et al. (2002)
Caldicellulosiruptor saccharolyticus	Sucrose	3.3	van Niel et al. (2002)
Enterobacter			
E. aerogenes	Glucose	1.0	Tanisho et al. (1983)
E. aerogenes	Sucrose	1.5	Tanisho et al. (1983)
E. aerogenes	Glucose	1.6	Tanisho et al. (1998)
E. aerogenes (mutant)	Glucose	3.4	Kumar et al. (2001)
E. cloacae	Glucose	2.2	Kumar and Das (2000)
E. cloacae	Sucrose	6.0	Kumar and Das (2000)
E. cloacae	Cellobiose	5.4	Kumar and Das (2000)
E. coli	Glucose	1.2	Ishikawa et al. (2006)
E. coli	Glucose	0.9	Blackwood et al. (1956)
E. coli	Glucose	2.0	Bisaillon et al. (2006)
E. coli	Glucose	2.0	Turcot et al. (2008)
Mixed Cultures	Cellulose	2.4	Ueno et al. (1995)
Mixed Cultures	Glucose	1.7	Chen, Kin & Chang (2001)
Mixed Cultures	Sucrose	3.4	Chen, Kin & Chang (2001)
Mixed Cultures	Hexose	1.7 to 2.5	Noike & Mizuno (2000)

1.5 Bacterial growth medium design and optimisation.

The main objective in the formation of a microbial culture medium is ensuring essential nutrients are present in suitable forms and concentrations. Media can be formulated as chemically defined or complex forms.

During the process of fermentation when carbon metabolism occurs high-energy phosphate compounds (e.g. ATP) are produced. The concentration of carbon source is significant in medium constituent formulation to ensure the synthesis of secondary metabolites. This is because certain carbon source concentrations may lead to repression of one or more enzymes used for the production of a secondary metabolite. This is known as catabolite repression or carbon source repression (Greasham & Herber, 1997).

Nitrogen is required to support the biosynthesis of some metabolites; this can be both primary and secondary. Some of the primary metabolites include purines, pyrimidines and amino acids. There are different types of nitrogen source; selection of a type nitrogen source may have the effect of a drift in pH during microbial growth (Greasham & Herber, 1997, chap. 3).

In order to satisfy the mineral requirements for microbial cells, fermentation media regularly include phosphate, calcium, magnesium, manganese, zinc sulphate, potassium and ferric ions. Phosphate is also a component of phospholipids, nucleic acids, constituent of high-energy compounds such as adenosine triphosphate, adenosine diphosphate and also a regulator of some enzymes of both primary and secondary metabolism (Martin, 1989). Growth of bacteria can be affected by pH and sometimes it affects the biosynthesis of primary and secondary metabolites. Medium optimisation is simply a search for the conditions that support the maximum growth and or synthesis of target microbial products. A review of different medium constituents that support growth in a culture medium and their effects on hydrogen productions are detailed below:

1.5.1 Carbon source requirements

The concentration of a carbon source is important during medium design to encourage the production of the targeted secondary metabolite. When the use of excessive carbon source concentrations is detrimental, repression of one or more enzymes needed for the synthesis of a secondary metabolite can occur (Greasham & Herber, 1997). It is known that the increase in carbon substrate concentration results in a metabolic pathway change during microbial fermentation (Yu & Fang, 2001). Microorganisms reduce carbon substrates to increase in cell density and for the production of metabolites as well as for energy source. ATP and other high-energy phosphate compounds are produced during carbon metabolism either by substrate level phosphorylation or oxidative phosphorylation. Low initial glucose concentration levels are important in the quality and hydrogen production rates during the fermentation. The effect of initial culture medium glucose concentrations during fermentation was researched by Alalayah et al. (2009). The highest yield of hydrogen was achieved when initial glucose concentration was 10 g/l, but there was a decrease with increasing glucose concentration.

Ding et al. (2009) reported that in co-cultures, the hydrogen yields and total volume of hydrogen produced were influenced by glucose concentration. The hydrogen yield and total hydrogen volume increased with increasing initial glucose concentration from 3 g/l to 6 g/l and 3 g/l to 12 g/l, respectively. When the initial glucose concentrations were increased from 6 g/l to 15 g/l and 12 g/l to 15 g/l there was a decrease in hydrogen yield and total hydrogen volume respectively. A maximum of 2.98 mol H_2/mol glucose was achieved at glucose concentration of 6 g/l and at this concentration the entirety of glucose was consumed. Concurrently, organic acid concentrations also increased and hydrogen yield decreased as glucose concentration was increased from 6 g/l to 15 g/l. This production of organic acids, results from the utilization of carbon substrates. Organic acids are responsible for pH decline, which is unfavorable to hydrogen production during photo or dark fermentation.

1.5.2 Influence of culture medium pH.

The pH of a culture medium is influential to microbial growth and yield of metabolic products in a fermentation process. Culture pH is an important influence in hydrogen production due to its effects on FeFe-hydrogenase enzyme activity, metabolic pathways changes, and influence on the duration of fermentation lag phase (Dabrock et al., 1992). Kapdan & Kargi (2006) reported

41

that low initial pH values of 4.0-4.5 are responsible for longer lag periods. Also, there is usually a decrease in lag time at high initial pH values of 9.0 with a lower yield of hydrogen production (Zhang, Liu & Fang, 2003). It has been reported that the effect of pH is as a result of the change in ionization state of the components in enzymatic reactions (Fabiano & Perego, 2002). Low culture medium pH during fermentation leads to a low balance of intracellular ATP, as ATP will be used to excrete H^+ from the cell, resulting in the inhibition of bacterial growth (Bowles & Ellefson, 1985).

1.5.3 Nitrogen source requirements

Microorganisms require nitrogen sources to support the biosynthesis of nitrogenous metabolites. These metabolites include primary (pyrimidines, purine and amino acids) and secondary metabolites. The primary metabolites include purines, pyrimidine and amino acids. Zhao et al., (2011) demonstrated that supplementing culture media with 4 different nitrogen sources (yeast extract, peptone, beef extract and urea), the medium supplemented with yeast extract exhibited the highest hydrogen evolution of 1417 ml H_2/l medium, hydrogen yield of 1.26 mol H_2/ mol glucose, CDW of 1.14 g/l, organic acids produced which resulted in the lowest final pH of 4.81. It was concluded that a suitable nitrogen source could facilitate cell growth and hydrogen evolution and suggested that a lower hydrogen yield in the presence of a single nitrogen source proved that combined nitrogen sources may be beneficial to improve bacteria cell growth as well as hydrogen production.

1.5.4 Carbon, Nitrogen and Phosphorus (C/N/P) ratios

Microorganisms need ideal nitrogen supplement as well as a source of carbon for metabolism during the process of fermentation. The carbon to nitrogen (C/N) ratio is a vital criterion in biological fermentation processes. In order to optimize hydrogen production from wastewater, an ideal C/N-ratio is needed in a mixed microflora. Lin & Lay (2004) investigated the effect of carbon/ nitrogen (C/N)-ratio on bio-hydrogen production when sucrose was utilized as a carbon substrate. It was noted that hydrogen production was dependent on the C/N-ratio influx. A maximum of 4.8 mol H_2/ mol-sucrose was attained at a C/N-ratio of 47 and there was a 500% increase in hydrogen yield compared to the initial medium formulation. Ideal formulation of C/N-ratio on hydrogen production was achieved by a by shift in the metabolic pathway.

Argun et al. (2008) reported on the effects of C/N and C/P ratio on hydrogen yield and hydrogen production rate during a dark fermentation process, utilizing wheat powder solution (WPS) as a carbon substrate. It was demonstrated that the yield of hydrogen produced increased with increasing C/N and C/P ratios. For high yield of hydrogen to be achieved a combination involving low nitrogen concentrations required low phosphorous concentrations as well.

1.5.5 Element requirements

In order to satisfy the mineral requirements for microbial cells, fermentation media regularly include phosphate, calcium, magnesium, manganese, zinc sulphate, potassium and ferric ions. Phosphate is also a component of phospholipids, nucleic acids, constituent of high-energy compounds such as adenosine triphosphate, adenosine diphosphate and also a regulator of some enzymes of both primary and secondary metabolism (Martin, 1989). They are also known as inorganic micronutrients or trace elements required in trace amounts as cofactors or activators for many enzymatic systems. However sulphate is also essential for biosynthesis of some secondary metabolites and sulphur containing amino acids (Greasham & Herber, 1997).

The effect of iron on the bacterial metabolism during the process of hydrogen production fermentation has been reported. Lee, Miyahara & Noike (2001) reported on iron-sulfur effect using anaerobic sewage microflora. They stated that iron sulphur affects protein functions primarily as an electron carrier and in metabolic changes involving pyruvate oxidation to acetyl coenzyme A, CO_2 and H_2. Iron could influence metabolic pathway changes and be involved in Fe-S and non Fe-S proteins functioning in the enzyme hydrogenase (Lee, et al., 2001). The addition of iron sulfate influences growth and the rate of hydrogen production, Alalayah et al. (2009) reported on the effect of iron on hydrogen production, at $FeSo4.7H_2O$ concentrations of 1-100 mg/l. It was concluded that the highest yield of hydrogen was observed at $FeSO_4.7H_2O$ concentration of 25 mg/l.

Lin & Lay (2005) reported that iron, magnesium, zinc and sodium, relate to the bacterial enzyme cofactor, transport processes and dehydrogenase. In their research it was concluded that hydrogen production reached a maximum value of 3.52 mol H_2 / mol of sucrose at 120 and 1000 mg /l of $MgCl_2$ and NaCl concentrations respectively.

The influences of Mg, Na and Fe on hydrogen production are dependent on their individual concentrations. Iron is an important element that forms ferredoxin and hydrogenase and has been reported by Das & Veziro˜glu (2001) as the most important trace metal influencing hydrogen production. The study by Lin & Lay (2005) proved that magnesium concentration variations readily influences hydrogen production. This also suggests that the right combination and concentrations of Fe and Mg are important for efficient hydrogen production.

The right medium constituents and concentrations are needed for an optimization process to enable the effective production of a target metabolite. This leads to medium optimization techniques, which is simply a search for the conditions that support the maximum growth and or synthesis of target microbial products.

1.6 Experimental design methods

Experimental design is a process carried out where certain factors are selected and consciously varied in a controlled condition to find their individual effects; this is often followed by the analysis of the experimental results. In an experimental design for optimization of hydrogen production, a number of factors and influences may be considered. Deciding to use either pure cultures or mixed cultures can be the first conundrum. Using mixed cultures to produce hydrogen, the hydrogen produced by the hydrogen-producing bacteria can be used and consumed by the hydrogen-consuming bacteria in the mixed cultures. Furthermore there can be competition for substrate, which may lead to substrate limitation for the hydrogen-producing bacteria. Therefore, in order to exploit hydrogen produced from a fermentative hydrogen production process, there needs to be a pretreatment to curb hydrogen consumption by the hydrogen-consuming bacteria whilst allowing the activities of the hydrogen-producing bacteria to prevail (Wang & Wan, 2008)

The number of the factors to be investigated at a time determines the experimental design and this can be classified into two categories: one-factor-at-a-time design and factorial design (Luftig & Jordan, 1998). Some examples of experimental designs include Taguchi design, Plackett and Burman design, central composite design and Box–Behnken design. Before the initiation of a process to optimize fermentative hydrogen production, the experimental design is very important, this is because the fermentation process is influenced by various factors for example the species of hydrogen producing bacteria, inorganic nutrients, substrates, and the environmental operational conditions of the bioreactors. Consequently a suitable experimental design can be explored to investigate the effects of different factors on the fermentation process to achieve a better understanding with an improved performance (Li & Fang, 2007).

1.6.1 One-factor-at-a-time design

This design investigates one factor, while making the levels of other factors constant. When investigating a factor the level of the factor to be investigated is changed to analyze its effects on a response.

One-factor-at-a-time design has two main downsides.

- It does not consider the interactions between different factors. This does not ensure that the optimal conditions identified by it are truly optimal, particularly if the interactions between different factors are significant. Kim et al. (2006) studied the effect of substrate concentration on hydrogen production using a continuous-flow stirred-tank reactor (CSTR). Sucrose was used as the main source of carbon. The effect of only one factor, which was sucrose concentration on fermentative hydrogen production was analyzed using the one-factor-at-a-time design, the interaction between sucrose concentration and other factors for example temperature were ignored.

- It is laborious and time consuming as it involves a large number of experiments especially when the numbers of factors involved are large (Kennedy & Krouse, 1999). Chittibabu, Nath and Das (2006) studied individually, the effects of initial substrate concentration, temperature inoculum size, initial medium pH, and dilution rate on the production of hydrogen, using the one-factor-at-a-time design, with approximately 30 experimental runs.

As one- factor-at-a-time design is easy to use, understand and analyze, it has been extensively applied to study the effects of numerous factors on the production of hydrogen by fermentative bacteria. Kim et al. (2006) studied the effects of different sucrose concentrations on the production of hydrogen by applying the design using one-factor-at-a-time technique. Results of this design were analyzed to illustrate the effects of sucrose concentrations on hydrogen production yield, hydrogen production rate and specific hydrogen production rate. They determined the optimum sucrose concentration for the hydrogen production process was 30 g COD/l.

1.6.2 Factorial design

This design investigates the effects of more than one factor at two or more levels. This experimental design involves combination of different factor levels, which enables it to illustrate the interactions between different factors. It is also a more efficient method to investigate a large number of factors. Factorial design can be classified into two categories: full factorial design and fractional factorial designs (Kennedy & Krouse, 1999).

In factorial designs there are coded factor level as well as actual factor levels. Coded factor levels present a uniform structure to research the effects of a factor in every experimental circumstance, whilst the actual factor levels is dependent on a specific factor to be investigated. However in factorial designs, factors are generally given as coded factor levels (Kuehl, 2000). Individual actual factor levels can be assigned to the matching coded factor level of a factorial design when applying the design. Coded or actual factor levels can be applied to derive the analysis or model fitting for a factorial design. However, in most cases, the coded factor level analysis is desirable, this is because using the coded factor level analysis, the model coefficients involved are dimensionless, and therefore directly comparable, as a result using the coded factor level is beneficial to distinguish the relative size of factor effects (Montgomery, 2005).

1.6.3 Full factorial design

In this type of design all factor level combinations are tested. In a full factorial design, every combination of each factor level is tested. Examples of factors are temperature, medium component, pH and inoculum. Full factorials provide cover for all analysis at the cost of a great number of experimental tests. In an un-replicated experiment for a full factorial design: two-factor designs are denoted by $a \times b$ and three-factor designs by $a \times b \times c$. All factors are tested at their individual levels for example in $a \times b$, the first factor is tested at a levels and the second at b levels; in $a \times b \times c$, a third factor is tested at c levels. This notation extends to a^n—a complete factorial for n factors, each at a levels. These designs provide the least detailed information on factor interactions. This is because when there are a great number of factors to be studied a great number of experimental runs are required. As a result of this a large experimental run is not economically and practically feasible (Luftig & Jordan, 1998).

For a full factorial design, the number of runs for n number of factors each at a levels is a^n. The two level designs is the most commonly applied full factorial design and when there are n numbers of factors it is denoted by 2^n (Kennedy & Krouse, 1999). Occasionally, a suitable polynomial model can be applied to illustrate the effects of the factors investigated on a response and then the response can be optimized if needed too. Therefore, all feasible combinations of factor levels in a full factorial design can be studied, this design has been used a large number of times to research on the effects of numerous factors simultaneously on the hydrogen production process by fermentative bacteria. An example of the full factorial design was applied by Chou, Wang, Huang and Lay (2008) with 24 experimental runs; they researched the effects of pH (at 4 levels) and stirring speed (at 6 levels) on the production of hydrogen by fermentation. Two second-order polynomial models were also formulated to show the effects produced by the two factors on hydrogen production yield and specific hydrogen production rate. It was determined that the optimum pH and agitation speed for hydrogen production by the fermentative bacteria were 6.0 and 120 rpm, respectively. Given that every feasible combination of each pH and agitation speed level was evaluated, the experimental design interactions concerning the two factors were depicted (Chou et al., 2008).

In a full factorial design, as the number of factors increase, the number of runs increases geometrically as well (Table 1.2). In other words in a 2 level experiment the number of run combinations grows according to the number of factors presented. Table 1.1 explains this relationship further.

Table 1.2 Example of a Full factorial design

Number of factors	Number of runs required
2	4
3	8
4	16
5	32
6	64
7	128
8	256

For example, Espinoza-Escalante et al. (2008) studied the effects of thermical treatment, alkalinization and sonication, each studied at 2 levels on the production of hydrogen by fermentation applying the full factorial design. With trials or runs studied at two level, if the effects of two factors were investigated using the full factorial design for hydrogen production by fermentation, 2^2 runs of experiment were needed, and if the effects of the three factors were investigated for hydrogen production by fermentation applying the full factorial design, 2^3 runs of experiment were needed. In many cases, as the effects of a great number of factors involved are to be investigated simultaneously, a large number of runs of experiment are needed. This will involve a large number of experiments, which is not economically and practically feasible (Luftig & Jordan, 1998).

In an experiment when using the control factor levels chosen, the full factorial design gives understanding about the factors, which control the experimental process, the size or extent of their effect and also the direction of their effect. This design also helps to decide if any of the factors involved in the design interact (positively or negatively) with any of the other factors in the design, which is critical in the medium optimization process. Application of this

51

design is also ideal for research and development and also for new product and development design. At the same time as it is important to know which factors actually influence or control experimental processes, similarly it is important to identify which factors have no noticeable effect on desired response. The main downside of the full factorial design, in an industrial environment, is the time and expense involved in applying this design in experiments.

1.6.4 Fractional factorial design

In a process when the number of experimental runs for a full factorial design is large, the required information can alternatively be obtained by using only a fraction of the full factorial design, this is referred to as fractional factorial design. The effects of certain factors on a response can be analyzed economically and practically using this design (Luftig & Jordan, 1998). In other words, fractional factorial design offers another method, if the number of experimental runs for a full factorial design is too large to be applicable. Taguchi design, Plackett and Burman design, central composite design and Box–Behnken designs are examples of fractional factorial designs.

1.6.4.1 Taguchi design

In this design the effect with two or more levels on a response can be investigated in a small number of runs using the orthogonal array. Using this method appropriately, Taguchi design may provide a powerful and efficient approach to find out the optimum conditions using the optimal combination of factor levels. Normally, with the support of range analysis, analysis of variance or analysis of signal-to-noise ratio, the important factors that have significant effects on a response are identified and most significant factor levels for a particular process can be identified from the pre-determined factor levels (Antony, 2006). Using a combination of orthogonal Latin squares; Taguchi formulated a different set of standard orthogonal arrays to be applied. An example of an orthogonal array developed for a 2 level factor is shown below.

Table 1.3 Example of an Orthogonal Array.

Trial Number	Factors						
	A	B	C	D	E	F	G
1	1	1	1	1	1	1	1
2	1	1	1	2	2	2	2
3	1	2	2	1	1	2	2
4	1	2	2	2	2	1	1
5	2	1	2	1	2	1	2
6	2	1	2	2	1	2	1
7	2	2	1	1	2	2	1
8	2	2	1	2	1	1	2

The array shown table 1.3 is applied to design the experiments with up to 7 factors with 2 levels. This array design is composed of 7 columns and 8 rows. Each row in the array signifies a trial run with factor levels denoted by the numbers in the row while the vertical columns relate to the experimental factors involved in the investigation. There are four level 1 and four level 2 conditions for each factor represented in the column. These factors combine in four possible ways for example two 2 level factors combine as (1,1), (1,2), (2,1), (2,2). A column is known to be orthogonal or balanced when two columns of an array show these combinations the same number of times. It is important to note that two columns of the orthogonal array above have the same number of combinations of (1,1), (1,2), (2,1), and (2,2). Hence the seven columns are orthogonal to each other.

Before the initiation of an experiment, it is critical for the design of the experiment to be accurate and ideal for the process. Therefore it is important to select an appropriate orthogonal array, put the factors to the right columns, and then explain the combinations each experiment. Each combination of factors in a particular experiment is called the trial conditions. In table 1.3, 8 trials are required to complete an experiment and also the level of each factor associated with each trial run. The experimental design is understood by reading the numerals 1 and 2 indicated in the rows of each trial run. This factorial experiment would need a total of 2^7 or 128 runs.

Analyzing the results from experiments carried out using the Taguchi method enable experimenters to achieve the following:

- To determine the optimum condition required for a product or an experimental process.
- To determine the influence of individual factors
- To evaluate the response of an effect at the optimum conditions

Investigating the main effect of each of the factors helps in the determination of the optimum conditions needed for the process study. The main effect reveals the trend of the influence of the experimental factors involved. The level of a factor, which produces the best effect, can be predicted by knowing if a high or low value produces a desired result.

Knowing the influence in an experiment contributed by individual factors is vital to choosing the type of control to be applied on a production process. The analysis of variance (ANOVA) is commonly used to know the amount contributed by each individual factor involved in the experimental process. However, the optimum conditions are not set out in numerous experiments carried out already because the orthogonal arrays depict only a small portion of all the potentials. By applying the Taguchi design, Lin and Lay (2005) investigated the effects of 13 nutrient concentrations on hydrogen production by fermentation. According to the analysis, they concluded that sodium, zinc, magnesium, and iron were essential trace metals influencing the production of hydrogen and also the best nutrient levels for the production of hydrogen were determined from the pre-determined factor levels. Though, the actual optimal factor levels may not be certain with the use of Taguchi design, this is because the actual optimal factor levels might be different from the related pre-determined factor levels (Antony, 2006).

1.6.4.2 Plackett and Burman design

Using this design it is known that there may be a large number of factors that influence a process, but only some factors have significant effects. Those factors that influence the process greatly are then given greater focus than those that influence the process slightly. Therefore, at the initial stages, identifying those factors with significant effect should be the first step. This method makes it possible to investigate an experimental run and identify those variables, with significant effects on the desired response from a larger number of likely variables (>5) (Greasham & Herber, 1997).

The number of runs for a Plackett and Burman design is equal to a multiple of 4. Each design in the Plackett and Burman method is a fraction of the two level factorial designs. "The maximum number of variables that can be evaluated in one design is equal to one less than the number of individual experiments" (Greasham & Herber, 1997). In other words the design can study up to n=N-1 factors in an experiments with N number of runs. There are usually unassigned variables known as dummy variables that are used to calculate the experimental error during analysis of experimental results. A number of replications are usually executed to estimate the experimental errors.

A first-order or degree polynomial model (Eq. 1.1) can be used to explain the effects of several factors, according to the experimental results derived from the Plackett–Burman design.

$$Y = \beta_0 + \sum \beta_i \chi_i \quad \text{............ (1.13)}$$

Y represents the response, β_0 is the constant or model intercept, β_i is the linear coefficient, and χ_i is the coded factor levels, which is the level of independent variable.

Significant factors can be identified from the Plackett and Burman design using the analysis of variance (Weuster-Botz, 2000). Pan, Fan, Xing, Hou and Zhang, (2008) studied the statistical optimization of process parameters involved in bio hydrogen production using glucose as its main substrate by *Clostridium* species Fanp2. They applied the Plackett and Burman design to investigate the effects of 8 factors on the production of hydrogen by fermentation. They identified 3 factors (glucose, phosphate buffer and vitamin solution) to have significant effects on the fermentation process according to the desired response. The desired response was the specific hydrogen production potential according to the analysis of the experimental results.

Furthermore Guo et al. (2009) investigated the optimization of culture conditions required to improve the hydrogen production by *Ethanoligenens harbinese* B49 using the response surface methodology. Six factors (nutrients) $FeSO_4.7H_2O$, $MgCl_2$, $CaCl_2.2H_2O$, $ZnSO_4.7H_2O$, NaCl and K_2HPO_4 were studied to investigate the significant ingredients influencing hydrogen production. Applying the Plackett and Burman design Fe^{2+}, Mg^{2+}, K_2HPO_4 and Ca^{2+} showed a positive effect for hydrogen production while NaCl and Zn^{2+} displayed a negative effect on hydrogen production. In conclusion Fe^{2+} and Mg^{2+} were identified as the significant nutrient salts for hydrogen production.

1.6.4.3 Method of steepest ascent

In a process optimization design, the best design is one that allows the optimization of a desired response by determining the best control settings for the individual factors involved. The basic goal in achieving this is to locate area of optimum.

In an experimental design the initial assumption of the optimal conditions for a bioprocess is not close to the actual optimum. Therefore, the second step in optimizing an experimental design is to determine the region of factor levels that yield an optimal set of conditions. The method is also a simple and economically efficient method provided to change the experimental region of a response in the direction of the maximum change approaching the optimum. In the case where, minimization of a response is required, then this method is known as the method of steepest descent.

However in the application of the Method of steepest ascent design a factorial or fractional factorial design is used initially to screen out irrelevant factors. An example of a fractional factorial design that can be used is the Placket and Burman design. These factors screened by the Plackett and Burman design can then be further researched using the Method of Steepest Ascent.

The Method of Steepest Ascent experimental procedures begins from the design centre of the factorial design and stops when no more improvement or enhancement can be attained from the effect of the response, which shows that the region of optimal response is in the region of that manipulated variable and its condition (Kuehl, 2000).

Pan et al. (2008) applied the Method of Steepest Ascent to study the statistical optimization of the process parameters involved in the production of hydrogen while utilizing glucose by *Clostridium* specie Fanp2. They used this method to investigate the region of factor levels that produced optimum conditions for the production of hydrogen by fermentation. They found the design centres of glucose, phosphate buffer and vitamin solution for further media optimization process (Box-Behnken design). The maximum bio hydrogen produced was 4083.2 ml H_2/l at the best region of glucose 22g/l, phosphate buffer 0.15 (M) and vitamin solution 15 ml/l.

Furthermore Long et al. (2010) investigated the statistical optimization of hydrogen production by fermentation using xylose as a carbon source, utilized by newly isolated *Enterobacter sp*. CN1. Centered on the results of the first regression analysis of Plackett and Burman design, the Method of the Steepest Ascent was applied to locate the accurate direction of changing variables. The volume of hydrogen was increased when this method was applied, increasing the concentration of xylose and $FeSO_4$ while decreasing the concentration of peptone helped to achieve this. The highest volume of hydrogen produced was 1075.09 ml H_2/l medium at concentrations of xylose 14 g/l, $FeSO_4$ 0.35 g/l and peptone 3.0 g/l. This suggested that the area of optimum response was close and these values were chosen for further optimization.

When the region of optimal response is identified by the use of the Method of Steepest Ascent, it is essential to illustrate the response in that region. Central composite design and Box–Behnken design are commonly used experimental designs for response surface methodology to assess a second-order polynomial estimate to a response in that region.

1.6.4.4 Central composite design (CCD)

The central composite design (CCD) was introduced and designed by Box and Wilson (1950). Central composite designs are composite designs fashioned from two-level factorials; this was done by adding extra points to evaluate and estimate curvature and interaction effects when applied in an experimental process. This design can be regarded as partial factorials with experimental factors at five levels, in other words it is also known as a five-level fractional factorial design (Box and Wilson, 1950). The number of experimental runs involved when applying the central composite design increases exponentially with the number of experimental factors involved in the process. For experiments with two factors, 16 experimental runs are ideal (Kennedy & Krouse, 1999).

After applying the Method of Steepest Ascent to identify the vicinity of optimum response, it is important to characterize the response in that vicinity. Central composite design and Box–Behnken design are experimental designs widely used for approximating second order response surfaces. (Park, Kim & Cho, 2008). Central composite design usually consists of a 2^n full factorial design, m central designs and 2 x n axial designs. The axial design is identical to the central design, but the central design uses levels either above the high level or below the low levels of the 2^n full factorial design (Kuehl, 2000).

Sittijunda and Reungsang (2012) studied the medium optimization criteria involved in the production of bio hydrogen from utilizing waste glycerol by mixed cultures of anaerobic thermophiles. The central composite design was used to optimize the medium composition influencing thermophilic biohydrogen production from utilizing glycerol. The medium components studied were urea concentration, waste glycerol concentration, disodium hydrogen phosphate (Na_2HPO_4) concentration and the quantity of endo-nutrient addition. Volume of maximum hydrogen predicted from the central composite design was 1470.19 mL H_2/L under optimum conditions. Batch fermentation experiments were carried out in triplicate to validate the predictions; however a total volume of 1502.84 mL/H_2/L of hydrogen was produced. This value was close to the predicted value and only differed by 2.22%. This result implied that the model derived from the CCD experiment is valid.

1.6.4.5 Box–Behnken design

This is a three-level fractional factorial design developed by Box and Behnken (Wang & Wan, 2009). The design is a blend of a two-level factorial design alongside an incomplete block design. In a particular block, a number of factors are set through all combinations for the factorial design, whilst all other factors are retained at the central levels. To describe this design, the Box-Behnken design can be applied for three independent variables and 13 experimental trials is shown in the table below.

Table 1.4 Box-Behnken experimental designs for three variables.

Experimental Trials	Independent Variables		
	X1	X2	X3
1	+	+	0
2	+	-	0
3	-	+	0
4	-	-	0
5	+	0	-
6	+	0	-
7	-	0	+
8	-	0	-
9	0	+	+
10	0	+	-
11	0	-	+
12	0	-	-
13	0	0	0

In table 1.4 each column represents an independent variable and each row represents an experimental trial. High, medium and low levels of each independent variable are denoted by +, 0 and – respectively. Values attributed to the levels of individual variables must be equally spaced in the experimental design. Experimental trials are usually carried out in a random order, in order to avoid bias in experimentation.

Pan et al. (2008) studied the effects of glucose, phosphate buffer and vitamin solution on fermentative hydrogen production using this design. They determined that glucose and vitamin solution and glucose and phosphate buffer had interactive effects on the production of hydrogen. The optimal conditions were glucose 23.75 g/l, phosphate buffer 0.159 mol/l and vitamin solution 13.3 mL/l.

Guo et al. (2009) studied the optimization of culture conditions to improve hydrogen production by *Ethanoligenens harbinense* B49. They applied the Plackett and Burman design to first screen out significant factors (Fe^{2+} and Mg^{2+}) then further optimized the combination of the selected factors to improve on the results. The significant variables and the sole carbon source (glucose) were used to further optimize the process by Box-Behnken method. A maximum hydrogen yield of 2.21 mol/mol glucose was predicted with the Box-Behnken method with concentrations of glucose, Fe^{2+} and Mg^{2+} to be 14.57 g/L, 177.28 mg/L and 691.98 mg/L, respectively. Batch experiments in triplicate were carried out to confirm this prediction. A maximum hydrogen yield of 2.20 mol/mol glucose was achieved based on the formulated optimized medium suggested. This result confirmed the practicability of this statistical optimization design. Furthermore Jo et al. (2008) investigated the statistical optimization of process variables for enhanced hydrogen production by *Clostridium tyrobutyricum* JM1. In this research they concluded that 3 key variables influenced hydrogen production rate. Then the variables (glucose concentration, pH and temperature) were further investigated to improve on the hydrogen production rate. The individual and mutual effects of three key process variables were also investigated in a batch-controlled system. This method was applied to investigate the effect of these variables and also to achieve optimum conditions for each variable or combination of variables that will produce the optimum result. Experimental results revealed that a maximum hydrogen production rate of 5089 ml H_2/g dry cells/h was achieved at glucose concentration of 102.08 mM, pH of 6.5 and a process temperature of 35°C. However all factors involved in the experimental process had significant influences on the specific hydrogen production rate.

The Box–Behnken design is an economical alternative to central composite design, this is because it uses less factor levels compared the central composite design as well as not containing extreme high or extreme low levels. After experimental trials are completed and responses determined in the Box-Behnken design, the coefficients of the quadratic polynomial models are calculated usually using regression analysis. Calculations are made by using computer programs such as JMP or Minitab software, that allows the generation of contours of responses against two independent variables (Greasham & Herber, 1997).

A visual representation by the surface plot and contour plot will show the response over a region of interesting factor levels. Furthermore, they will signify the sensitivity of the response to the change of each factor levels and also to the degree the factors interplay as response is affected. According to the analysis of variance, variables, which have significant effects on the response, are determined and also optimal levels can be determined.

Strict and facultative anaerobes which produce hydrogen through the fermentative hydrogen production process are beneficial and are better potentials for hydrogen production. This is because of the higher production rates of bio-hydrogen compared to the bio-hydrogen production through bio-photolysis and photo fermentation. In addition there are no inhibitory effect of oxygen on FeFe-hydrogenase under anoxic or anaerobic environments

(Tanisho & Ishiwata, 1994). Bio-hydrogen can be produced with or without the presence of light energy with the fermentative bacteria when compared to the photosynthetic bacteria. It can also utilize a variety of substrates as well as waste products (Nandi & Sengupta, 1998). Clostridia and Enteric bacteria, like *Clostridium butyricum* and *Enterobacter aerogenes* are known to produce hydrogen at high volumes and yields (Nath & Das, 2004). Facultative anaerobes like *E. aerogenes*, rapidly consume oxygen and resume the activity of the enzyme FeFe-hydrogenase under anoxic environments. This is different to strict anaerobes, which prove to be very sensitive to the presence of oxygen. Consequently, feed medium for this facultative anaerobe does not need to be degassed during a continuous production of hydrogen (Yokoi et al, 1995). Research investigations involving the optimization of bio-hydrogen production by *E. aerogenes* have been carried out (Jo et al, 2008) whilst studies investigating *C. beijerinkii* and *C. paraputrificum* for bio-hydrogen production have been rarely reported. The optimization of hydrogen yield from utilizing chitin and N-acetylglucosamine by *C. paraputrificum* has been performed as well (Dweirra et al. 2001). In addition, there is a shortage in the literature presently investigating the statistical optimization of medium constituent variables on biogas (bio-hydrogen) production by pure clostridial species. Futhermore, nothing has been reported about the optimization of biogas production from utilising N-acetylglucosamine by *C. beijerinckii* and *C. paraputrificum* through statistically designed methods.

1.7 Knowledge gaps

1) It is known that *C. paraputrificum* can utilize N-acetylglucosamine as a carbon substrate (Dwierra et al., 2000) however, the ability of *C. beijerinkeii* to utilize N-acetylglucosamine has not been reported.

2) The Placket and burman method has not been conducted to optimize and identify significant medium constituents that influence biogas production by *C. beijerinckii* and *C. paraputrificum* when utilizing N-acetylglucosamine and the main carbon substrate.

3) It is known that cysteine acts as a reducing agent, to improve the culture medium environment for growth of anaerobic bacteria however its effect on the metabolic production of organic acids during fermentation has not been reported. The effect of cysteine on the biochemical system will be explored to study its influence on cumulative biogas production.

4) There is no report, identifying the region of optimum medium formulation for maximum cumulative biogas volume by employment of the Method of Steepest Ascent for *C. beijerinckii* and *C. paraputrificum* cultures, utilizing N-acetylglucosamine.

5) Interaction effects amongst medium constituents and optimal medium formulation for maximum cumulative biogas volume have not been achieved when N-acetylglucosamine was utilized by *C. beijerinckii* and *C. paraputrificum* as a carbon source.

66

1.8 General introduction, aims and objectives.

Utilization of fossil fuels has caused an increase in CO_2 concentration, which is well known to be a main cause of pollution and of climate change. The exhaustion of fossil fuel resources will eventually result in increasingly severe energy shortages in the future. One promising solution to the exhaustion of fossil fuel resources and also to reduce environmental pollution is hydrogen.

Hydrogen is also well known as a non-polluting ideal fuel because water, which is non-toxic, is the sole product of combustion (Suzuki, 1982). Production of hydrogen by biological means is usually carried out at room temperature and a pressure of 1 atmosphere, and so the process is less energy demanding as compared with electrolytic and thermochemical processes.

One possible source of waste that could be recycled as a raw material for H_2 production is marine waste like crab, shrimp and lobster shells which are made up of Chitin. Chitin is the second most abundant polymer in the world after cellulose (Watson, Zinyowera & Moss, 1998). In most processes for hydrogen production, glucose (Masset et al., 2010) and cellulose (Levin et al., 2006) have been used as carbon substrates, continuous use of these carbon sources will eventually lead to food shortages.

Methods to utilise chitin or its monomer, N-acetyl glucosamine to produce H_2 are at an early stage of development (Evvyernie, et al. 2001). The statistical optimisation of medium constituents using N-acetylglucosamine as a substrate by *C. beijerinckii* and *C. paraputrificum* has not been reported.

The use of environmental and nutritional factors to optimize fermentation conditions, are important for application of bioprocesses. To optimize these processes using conventional techniques, for example the one factor at a time method, is time consuming and laborious (Greasham & Herber, 1997). Statistical design of experiments has become popular for medium optimization in industry since this allows the investigator to assess more than one factor at a time. This method is a time saving and effective method for the screening significant factors of a multivariable system. This has also been used extensively (Liu & Wang, 2007).

The hydrogen-producing Fanp3 strain of *Clostridium beijerinckii* has the ability to utilise a wide variety of carbon and nitrogen sources, which suggests that it has a high potential to convert various waste materials to valuable energy-yielding products like hydrogen. The *C. beijerinckii* Fanp3 maximum hydrogen yield was reported to be 2.52 mol H_2/mol glucose at a production rate of 39 ml H_2/ g of glucose/ h (Chun-Mei, Yao-Ting, Pan & Hong-Wei, 2008). *Clostridium paraputrifiicum* bas been reported to have the ability to utilize N-acetylglucosamine as a carbon source for hydrogen production. *C. paraputrificum*, strain M-21, produced 2.2 and 1.5 mol of H_2 gas from 1 mole of N-acetylglucosamine and ball milled chitin (equivalent to 1 mole of GlcNAc), respectively. (Evvyernie et al., 2001). These results suggest that *C. beijerinckii* and *C. paraputrificum* are ideal candidate organisms for biological hydrogen production from biomass waste. In order to reveal the potential for hydrogen production by these clostridial species, medium optimization was performed using N-acetylglucosamine as the major carbon source.

In the present work, media preparation methods and constraints, were analyzed. The following experimental design strategy for optimizing *C. biejerinckii* and *C. paraputrificum* in a fermentative biogas production process is highly recommended. A Plackett and Burman design was used to screen for the significant factors in a fermentative hydrogen production process. The Method of Steepest Ascent will be applied to approach the vicinity of the optimal conditions. The Box–Behnken design for response surface methodology will be applied thereafter to estimate the relationship between a response and the key factors at the vicinity of optimum and then find the optimal conditions.

1.8.1 Aims

- To determine the optimal medium environmental conditions, for the production of cumulative biogas (bio-hydrogen) during fermentation of GlcNAc by *C. beijerinckii* and *C. paraputrificum.*

1.8.2 Objectives

- To determine the ability of *C. beijerinckii* to metabolise the monomer GlcNAc as the sole carbon source.
- To use the Plackett and Burman method to identify medium constituent variables which significantly influence total volume of biogas produced during N-acetylglucosamine fermentation by *Clostridium beijerinckii*

NCTC 13035 and *Clostridium paraputrificum* NCTC Y1833.

- To investigate the formation of products and the optimum concentration of L-cysteine.HCl.H$_2$O within the growth medium for *Clostridium beijerinckii* to produce the highest cumulative biogas volume.

- To investigate the relationship between the metabolic shifts of organic acids produced and their individual concentrations on total volume of biogas produced during L-cysteine optimization.

- To determine the medium component conditions that yield optimal biogas produced by *C. beijerinckii* and *C. paraputrificum* using the variables identified by the Plackett and Burman design experiments. This is done by varying the medium constituents along the vicinity of optimum influence to produce the maximum cumulative biogas.

- To determine the optimum levels of the various medium constituents required for the maximum total volume of biogas to be produced by *Clostridium beijerinckii* and *Clostridium paraputrificum*.

- To study the interactive effects of the medium constituents tested in the *C. beijerinckii* and *C. paraputrificum* cultures.

Chapter 2
General Materials and Methods

Chapter 2: General Materials and Methods

This chapter describes the methods used to investigate the optimization techniques and analysis of the experimental results.

2.1.1 Bacterial strains and revival of bacteria cultures.

Clostridium beijerinckii NCTC 13035 and *Clostridium paraputrificum* NCTC Y1833 were supplied as freeze dried ampoules by the culture collection of the School of Life Sciences, University of Hertfordshire.

The revival / culturing medium for *Clostridium beijerinckii* NCTC 13035 consisted of N-acetylglucosamine 10g/l, bacteriological Pepetone 3 g/l, yeast extract 2 g/l, buffer capacity of 0.2M at a pH of 6.5, L-cysteine.HCl 1 g/l, $FeSO_4.7H_2O$ 0.1 g/l, $MgCl_2$ 0.213 g/l. The revival medium for *Clostridium paraputrificum* NCTC Y1833, consisted of N-acetylglucosamine 10g/l, bacteriological peptone 1 g/l, yeast extract 4.5 g/l, KH_2PO_4 3g/l, K_2HPO_4 5.8 g/l, $(NH4)_2SO_4$ 2.6 g/l, L-cysteine.HCl 1g/l, 3-(N-morpholino)propanesulfonic acid (MOPS), 10 g/l, and Na_2CO3 2.5 g/l. Steps for revival are stated below:

- The ampoule was sprayed with 70% ethanol and was aseptically transferred into a sterile laminar flow cabinet.

- When the ethanol had evaporated, the ampoule was cut open, revival medium was transferred with a pipette into the ampoule to refresh the Clostridium bacteria and the culture was then transferred into 100 ml culturing broth.

- This culture was incubated anaerobically at 37°C for 48 hours.

2.1.2 Bacterial storage conditions

After revival, stock cultures of each individual bacterium were stored in both 10 ml of 2.4% Difco bacto thioglycolate medium without dextrose, and bacteriological cryopreservation beads (Lab M D530) stored at -20°C. The thioglycolate medium (Difco) without dextrose contained Difco-bacto yeast (5 g/l), bacto casitone (15 g/l), L-cysteine (0.25 g/l), NaCl (2.5 g/l) thioglycollic acid (0.3 ml), bacto agar (0.75 g/l) and methylene blue (0.002 g/l). This medium was boiled to dissolve, mixed thoroughly and then 10 ml volumes of the medium were dispensed into test tubes which were then capped and sterilised by autoclaving at 121°C for 15 minutes. The test tubes were allowed to cool down and then inoculated aseptically with the individual stock cultures of the clostridial species used. The culture was incubated anaerobically at 37°C for 48 h and then stored at 4°C anaerobically until use.

2.1.3 Confirmation of purity of Clostridia culture.

Gram staining was used as a quick and simple method to determine that the putative clostridial strains were Gram-positive rods. They were observed under a microscope (Nikon Eclipse E100 Biological Microscope). *Clostridium beijerinckii* and *C. paraputrificum* form round oval endospores.

2.1.4 Media and Cultivation

Just before use, the stock cultures were incubated at 37°C for 24hrs. These stock cultures were used for all of the batch culture experiments. Prior to the batch culture growth, the *C. beijerinckii* and *C. paraputrificum* were revived in pre-culture medium. The pre-culture medium contained the following medium composition N-acetylglucosamine 10g/l, peptone 3g/l, yeast extract 1g/l, Buffering capacity (KH_2PO_4 and Na_2HPO_4) of 0.2M at a pH of 6.5, L-cysteine.HCl 0.5 g/l, $FeSO_4.7H_2O$ 0.1 g/l and $MgCl_2$ 0.21 g/l. N-acetyl glucosamine and the $FeSO_4.7H_2O$ were filter sterilised separately and mixed with the other medium components aseptically thereby making up the final culture medium.

The inoculum volume was 10%, with anaerobic incubation at 37°C and agitation of 120 rpm. The pre-culture experiments were conducted in a 250ml conical flask with a total volume of 100 ml. These flasks were inoculated separately with each of the clostridial strains from the stock thioglycolate medium cultures and then purged with oxygen free nitrogen gas for 10 minutes to displace the dissolved oxygen. This culture was grown for 2 days. The equipment for this process is shown in figure 2.1.

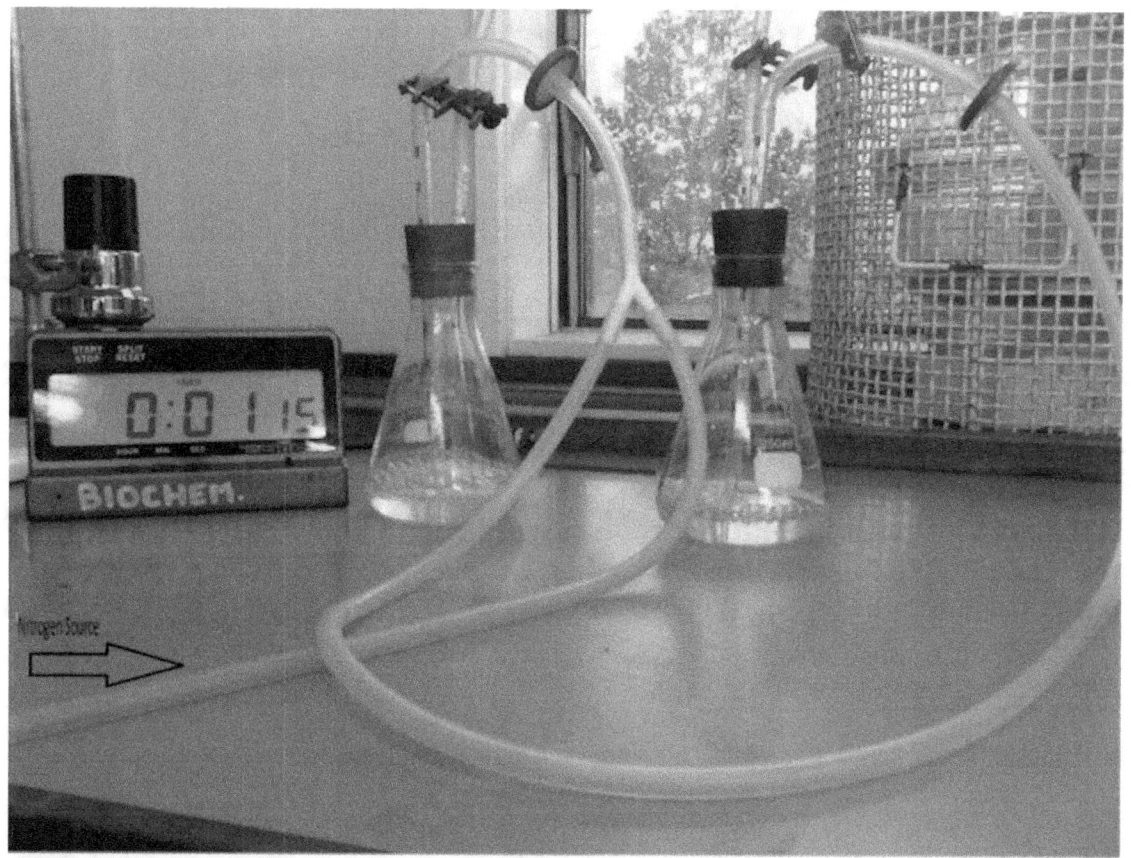

Figure 2.1 Purging of *Clostridium* growth media with oxygen-free-nitrogen to permit growth of the clostridial cultures.

2.1.5 Starter cultures.

Starter cultures were cultivated in 250 ml conical flasks. Starter medium contained N-acetyl glucosamine 10g/l, peptone 3g/l, yeast extract 1g/l, Buffering capacity (KH_2PO_4 and Na_2HPO_4) of 0.2M at a pH of 6.5, L-cysteine.HCl 0.5 g/l, $FeSO_4.7H_2O$ 0.1 g/l and $MgCl_2$ 0.21 g/l. Starter cultures were inoculated with 10% pre-culture. The total volume was 50ml.

The starter cultures were made anaerobic by purging with oxygen free nitrogen gas for 10 minutes to displace the dissolved oxygen (fig 2.1). The culture was incubated for 42 h at an agitation rate of 120 rpm at 37°C.

2.1.6 Batch cultures to check ability of *Clostridium beijerinckii* to utilise N-acetylglucosamine.

Batch culture experiments were performed in a 250 ml conical flask with a working volume of 100 ml to determine the ability of the *C. beijerinckii* to metabolise N-acetylglucosamine as their main carbon source. The medium proposed by Chun-Mei et al. (2008) was used; glucose was substituted with N-acetylglucosamine. Different control measures were used to check the ability of the bacteria to utilise N-acetylglucosamine. Positive and negative controls were used to check for biomass concentration, pH and sugar utilisation. In the positive control glucose was used as the major carbon source whilst in the negative control there was no added sugar. They were cultured in anaerobic jars for 6 days. To determine the ability of the organisms to use each of the nutrient supplies, the biomass concentration achieved at the end of the fermentation period in the control and experimental cultures was determined by checking the optical absorbance of final biomass concentration as explained in section 2.3.1.

2.2 Preparation methods for Plackett and Burman experiments.

2.2.1 Methods to maintain a constant Clostridium inoculum concentration.

It is important to always start the cultivation of these clostridia bacteria with a consistently high biomass concentration, in order to reduce the lag phase. A volume of 10 ml of inoculum from the thioglycolate medium culture was used to inoculate 90 ml of clostridial medium in a 250 ml conical flask. The medium was sparged with nitrogen gas to drive off the dissolved oxygen, for a minimum of 10 minutes, and then the flask placed in an anaerobic jar. The O_2 in the anaerobic jar was further sparged with N_2 gas for 2 minutes before being incubated at 37°C. This was incubated for 42 h in a shaker at 120 rpm. The method used to achieve a constant biomass concentration is explained in figure 2.2.

After 42 h of innoculation 10% of the initial culture was sub-cultured into another *Clostridium* medium, which was incubated for 42 h to produce a starter inoculation culture. This was conducted to scale up the volume of inoculum for experimental use and also to maintain a consistent inoculum.

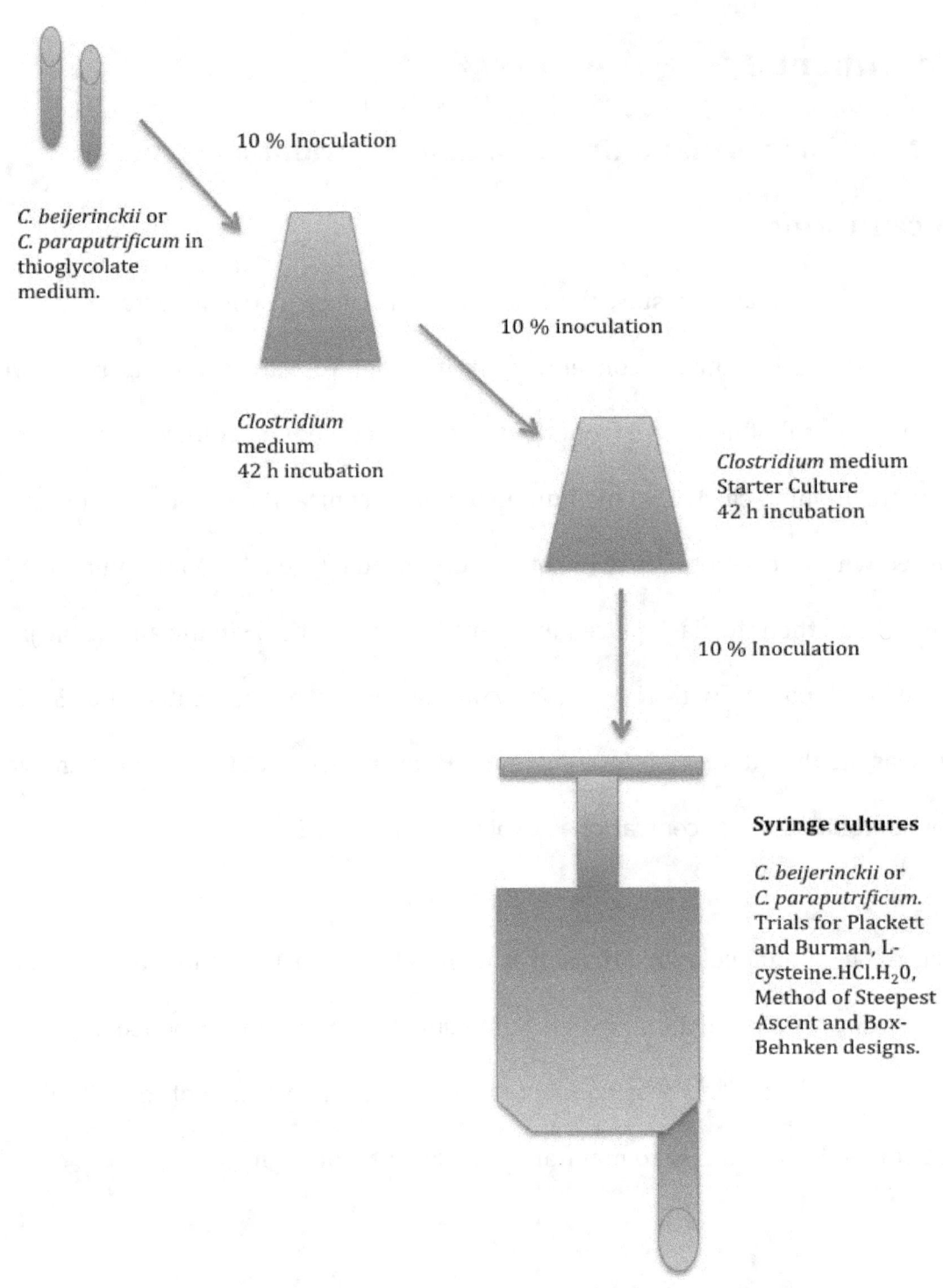

10 % Inoculation

C. beijerinckii or
C. paraputrificum in
thioglycolate
medium.

10 % inoculation

Clostridium
medium
42 h incubation

Clostridium medium
Starter Culture
42 h incubation

10 % Inoculation

Syringe cultures

C. beijerinckii or
C. paraputrificum.
Trials for Plackett
and Burman, L-
cysteine.HCl.H$_2$0,
Method of Steepest
Ascent and Box-
Behnken designs.

Figure 2.2 Diagram illustrating experimental method for preparing all optimisation trial designs in syringes.

2.2.2 Batch preparation methods for all statistical syringe culture experiments.

A total of 5 ml making a total of 50 ml of culture was used as the inoculum for all the experiments to determine the optimum medium composition by statistical method. These experiments include the Plackett and Burman design, one-factor cysteine experiments, Method of steepest ascent and the Box-Behnken method as described in subsequent chapters. Each volume of inoculum was centrifuged for 10 minutes at 5000 rpm, the supernatant was removed aseptically and the cells were re-suspended in 5 ml of the appropriate designed-medium made for the fermentation experiment. The re-suspended cells were used to inoculate 45 ml of the appropriate individual trial medium. A total of 10ml of inoculated medium was aseptically transferred into a sterile 50 ml Terumo syringe (Terumo Corporation Europe N.V., 3001 Leuren, Belgium).

These syringes were incubated at 37°C in an orbital shaker (Model G25 incubator shaker, New Brunswick Scientific co.inc Edison, New Jersey, U.S.A) at a rotational speed of 120 rpm. The distance the plunger of the calibrated syringe moved, as a result of the biogas produced indicated the total volume of biogas and were standardised to ml/l. All experiments were carried out independently in triplicate.

2.3 Equipment Calibration and measurement

2.3.1 Culture Optical Density measurements.

Samples were collected from the cultures immediately after inoculation and after each fermentation period. The density of bacterial culture was determined using a spectrophotometer (Cecil Ce2021 2000 series) by measuring the absorbance at 560 nm using the corresponding uninoculated medium as a blank.

2.3.2 Cell concentration analysis

Samples were collected from the cultures immediately after inoculation and after each fermentation period. The density of bacterial culture was determined by measuring the absorbance at 560 nm using the corresponding uninoculated medium as a blank. Subsequently the cell concentration was determined by converting the optical absorbance of the culture broth at 560 nm to dry cell weight per litre of individual culture broth by referencing to a standard curve that had previously been prepared.

To achieve a dry cell weight 47 mm glass microfiber filters (Whatman Cat number 1820-047) were dried at 65°C for a minimum of 6 hours until the weight was constant. Broth culture samples in different dilutions were filtered through and then the filters were dried in a 65°C oven until the weight was constant. Cell biomass was determined as the difference in mass between the filter without cells and the filter with a sample of cells.

2.3.3 Calibration of the bioreactor temperature probe, Bench pH Meter, Bioreactor pH – electrode and dissolved oxygen (pO$_2$) - electrode

Calibration of the temperature probe was a single point calibration conducted at ambient temperature according to the procedure specified by the manufacturer (B.Braun biotech international biostat B).

The bench pH meter (Mettler Toledo) was calibrated every day prior to use. This was conducted with a 2-point calibration with buffers of pH 4 and 7. The pH meter displayed the electrode slope value at the end of the calibration process. Note that on any occasion when the slope value was less than 95% the electrode was cleaned and calibration process was repeated again. The characteristics of pH electrodes normally change with time due to electrode coating and aging. It is usual to have slope values below 100% than above 100%. Slope values for electrodes in good working conditions are usually between 95% and 102%.

The bioreactor pH-electrode was calibrated using a two-point calibration, with buffers of pH 4 and 7, which determines the electrode parameters zero (pH 7) drift and slope using buffer solutions. The calibration of the bioreactor pH electrode was conducted before it was attached to the culture vessel. The calibration of the dissolved oxygen electrode was reported in percentage oxygen saturation, % pO$_2$. The calibration was conducted using a two-point calibration. The dissolved oxygen electrode was calibrated after sterilisation *in situ* in the bioreactor culture vessel.

2.3.4 Post sterilization checks and calibrations.

The culture vessel and all its working accessories were autoclaved at a temperature of 121°C for 15 minutes. The nitrogen supply for the bioreactor was fed into the culture vessel via a sterile membrane filter and sparged through a culture medium. During the supply of nitrogen the dissolved oxygen value was monitored continiously and checked if the value reduced slowly within 15 minutes to ensure that there was no damage to the PO_2 electrode.

When the dissolved oxygen value reduced and stabilised at a value of $0\pm5\%$ the nitrogen tap was closed to stop the supply of nitrogen. The culture medium was supplied with air through a filter vessel to also monitor the impact on dissolved oxygen concentration. The dissolved oxygen value should go up within 2 minutes. Once the dissolved oxygen value increased and was stable the air supply was turned off. At this point the post sterilisation check was complete.

2.3.5 Sterility Test

A fermentation run without inoculation was conducted. All necessary components were connected to the culture vessel and the necessary operating conditions for the fermentation run adhered too. The process was allowed to run for a 24 h test period. This duration is sufficient to detect microbial contamination. A sample was taken out from the vessel and checked under the microscope for contamination. If the sterility test is successful and there were no contaminants the fermentation run was allowed to continue as planned.

2.3.6 Pre-inoculation checks

The bioreactor stirrer impeller was set at 150 rpm and temperature set to 37°C. The sample port was sprayed with 70% ethanol and then allowed to dry. About 5 to 10 ml of culture medium was withdrawn aseptically from the bioreactor into the sample bottle.

The pH of the sample was measured with the offline pH meter (Bench pH meter). The online pH (Bioreactor pH) and the offline pH readings were compared and adjustments made to the online pH when necessary. If the pH difference was up to ± 0.1 the online pH value was adjusted to the offline value.

Steps were taken to ensure that the temperature of the medium reached the set values before recalibrating the pH with the offline value and also before inoculation.

2.3.7 Adding sterile solutions (N-acetylglucosamine and FeSO$_4$.7H$_2$O) during the fermentation run.

In certain fermentation processes, it is often necessary to introduce additional nutrient solutions or special essential nutrients. In this case N-acetylglucosamine and FeSO$_4$.7H$_2$O were filter sterilised (Using Whatman 45 mm glass microfiber filters) as they were added into the culture media in the bioreactor vessel. This was because these components are sensitive and cannot be sterilised by heat for this fermentation process. The medium component solutions made were transferred from a sterilised storage bottle into a sterile syringe then transferred to the culture vessel aseptically using a sterile filter.

2.3.8 Inoculating the Culture Vessel

The dissolved oxygen electrode calibration was carried out as described in section 2.3.3 and then sparged with nitrogen after calibration to a dissolved oxygen concentration value of 0%. The temperature of the medium in the bioreactor vessel was then allowed to stabilize at 37°C before inoculation. The culture medium was then inoculated with the starter culture. Inoculation of the culture vessel was performed using a sterile syringe. The inoculant (100 ml) was 10% of the total media volume (1litre).

2.3.9 Biogas measurement

The distance the plunger of the calibrated syringe moved, as a result of the biogas produced indicated the total volume of biogas and were standardised to ml/l. All experiments were carried out independently in triplicate.

2.4 Analytical methods

2.4.1 HPLC analysis.

Quantification of organic metabolites and residual N-acetylglucosamine was performed by HPLC. Samples were passed through an Aminex HPX-87H ion exchange column (300 mm by 7.8 mm Bio rad HPLC organic acid column) attached to a refractive index detector (Showa Denkos shodex RI-101), using an ultimate 3000 Dionex HPLC pump connected to a computer equipped with Chromeleon chromatography management software version 6.6. The software was used for quantification of organic metabolites and residual GlcNAc throughout the experiment. The mobile phase was 0.005M H_2SO_4 with a flow rate of 0.6 ml/min, pressure of 60 bars and a temperature of 35°C. The organic acid and the residual GlcNAc concentrations in samples were determined with reference to pre-prepared standard curves. The mean values were calculated for 3 replicates and standard deviations quantified.

2.4.2 HPLC standard solutions

N-acetylglucosamine (Oxoid), lactate (Sigma, 71716), formate (Sigma 17841), acetate (BDH 102363P), propionate (Sigma P1880), butyrate (Sigma, B5887) and ethanol (Sigma-Aldrich E7023) were used as standards for the sugar and organic metabolites. A stock solution containing a mixture of the GlcNAc and the metabolic end product standards (lactate, formate, acetate, ethanol, propionate and butyrate) were prepared as shown in table 2.1.

Table 2.1 Concentrations of HPLC standards.

HPLC Standards	Concentration of standards (g/l)
N-acetylglucosamine	15
Lactate	10
Formate	10
Acetate	10
Ethanol	5
Butyrate	5
Propionate	10

g/l = grams per litre.

A standard curve was prepared using the following concentrations 100%, 50%, 12.5%, 3.13%, 1.56% and 0.78% of the stock standard solution.

2.4.3 HPLC of metabolites and preparation.

Culture samples were transferred to separate sterile eppendorff tubes and then centrifuged in a microfuge (MSE Micro Centaur, MSE, UK) at 13000 rpm for 5 minutes. The supernatants were collected and analysed, immediately or stored for future use at 4°C. Samples were stored for up to 2 weeks at -20°C.

The concentrations of the organic metabolites and residual GlcNAc were determined in triplicate. Mean and standard deviation values were calculated.

Chapter 3

Identifying Significant Medium Constituents Using Plackett and Burman Method

3.1.1 Background

Statistical experimental designs are important techniques used for a wide range of practical work and to obtain reasonable and clear results cost-effectively. The Plackett and Burman design is an effective technique used to optimize culture medium constituents (Pan et al., 2008) and can be used to identify medium constituent factors, which significantly control hydrogen production. The constituents that have the most significant influence on the process are given more attention in medium design compared with those that have little influence. Pan et al. (2008) applied the Plackett and Burman design to study the effects of 8 culture medium constituents on the production of bio-hydrogen by *Clostridium species* Fanp2. They concluded that 3 of the medium constituents (glucose, phosphate buffer and vitamin solution) had significant effects on the production of bio-hydrogen, whilst the others did not. Bakonyi et al. (2011) also applied the Plackett and Burman experimental design to optimize the production of hydrogen by *E. coli* (XL1-BLUE). This process is dependent on the membrane-boundary formate-hydrogen lyase (FHL) enzyme complex. When analyzed, the results from the Plackett and Burman method indicated that out of the several medium constituents tested, only the formate concentration had a significant effect on the production of hydrogen. Furthermore the optimization of process variables was investigated for the production of hydrogen by *Enterobacter aerogenes* utilizing glucose. Three important variables were selected (glucose, initial pH and ferric chloride) and were subsequently identified as significantly influencing hydrogen production (Karthic, Shiny & Naveenji, 2012).

3.1.2 Objectives

- To use the Plackett and Burman method to identify medium constituent variables which significantly influence total volume of biogas produced during N-acetylglucosamine fermentation by *Clostridium beijerinckii* NCTC 13035 and *Clostridium paraputrificum* NCTC Y1833.

3.2.0 Materials and Methods

The bacterial strains used in this study were *C. beijerinckii* NCTC 13035 and *C. paraputrificum* NCTC Y1833. The storage conditions and revival were as described in section 2.1.2 and 2.1.1. Starter cultures were as described in section 2.1.5 and were used as the initial inoculum for this experiment. The initial inoculums volume used was 10% of the volume of the culture medium used. The bacterial cell concentration of the culture was determined as described in section 2.3.2. Metabolic products were analyzed and prepared as described in section 2.4.3. The medium design experiments were carried out in triplicate.

3.2.1 Ability of *C. beijerinckii* and *C. paraputrificum* to utilize N-acetylglucosamine as a carbon source.

The ability of the bacteria to utilize N-acetylglucosamine was determined because it is not currently known if *C. beijerinckii* is capable of metabolizing N-acetylglucosamine as its main carbon source. On the other hand, there are reports that *C. paraputrifucum* has the ability to utilize N-acetlylglucosamine (Evvyernie et al., 2000). The same procedure was carried out to determine the ability of *C. beijerinckii* as well as *C. paraputrifucum* to metabolise N-acetylglucosamine over a period of 15 days. The methods were as described in section 3.3.3.

3.2.2 Methods to check for constituents causing browning and precipitates after autoclaving.

Some medium constituents degrade or cause browning during autoclaving, therefore different constituent combinations were explored to identify those that caused browning and precipitate formation in the medium formulations being tested. Selected medium constituents tryptone, peptone, L-cysteine.HCl.H_2O, $FeSO_4.7H_2O$ and $MgCl_2$ were used in different combinations, but all at a concentration of 10 g/l. They were all autoclaved separately in 20 ml universal bottles. Some medium components were mixed and then autoclaved as well. This experiment was undertaken to achieve the correct medium formulation and combination, free from precipitates that could affect constituent concentrations, in preparation for the Plackett and Burman method.

3.2.3 Maintaining a constant Clostridium inoculum concentration.

It is necessary to initiate the cultivation of *C. beijerinckii* and *C. paraputrificum* with a high biomass concentration always, so as to reduce the lag phase of the fermentation process. Furthermore, it was necessary to maintain a constant

90

inoculum size for the Plackett and Burman method, since inoculum size was one of the dummy variables used in the experimental design. Each of the dummy variables used in the Plackett and Burman methods was kept at a constant level. The steps employed to maintain a constant inoculum size are described in section 2.2.1.

3.2.4 Experimental designs.

Steps for the culturing the Clostridia are described in section 2.2.2. However *C. beijerinckii* were cultured for 18 days in a 10 ml volume of culture medium held within a 20 ml syringe whilst *C. paraputrificum* were cultured for 18 days in a culture medium volume of 10 ml held within a 50 ml syringe. All experiments were carried out independently in triplicate. Initial carbon source concentrations were based on ranges quoted in the literature for studies on optimization as detailed in table 3.1.

Table 3.1. Ranges of initial carbon source concentrations used in studies.

Organisms Used	Carbon Substrate	Moles of H_2/ Mole of substrate	References
Ethanoligenes harbinense	10 g/l, 15g/l glucose	2.21	Guo et al. (2009)
Microflora	6 g/l glucose	2.98	Ding et al. (2009)
Clostridium specie	10 g/l, 15g/l glucose	2.53	Pan et al. (2008)
Clostridium butyricum	15.7 g/l glucose	2.2	Chong et al. (2009)
Clostridium beijerinckii	5 g/l, 10 g/l glucose	2.4 - 2.52	Pan, Yao & Houa (2008)
Clostridium paraputrificum	10 g/l N-acetylglucosamine	1.9	Evvyernie et al. (2000)

Table 3.2 shows the medium composition at its high and low design level for the culturing of *C. beijerinckii*.

Table 3.2 Levels of medium component variables for the culture of *C. beijerinckii* using Placket-Burman design.

Code	Variables	Low level (-)	High level (+)
X1	N-acetylglucosamine (g/l)	10	15
X2	Peptone (g/l)	3	4
X3	Yeast extract (g/l)	1	2
X4	Buffer capacity (M)	0.2	0.25
X5	L-cysteine.HCl (g/l)	0.5	1
X6	$FeSO_4.7H_2O$ (g/l)	0.1	0.4
X7	MgCl2 (g/l)	0.214	0.427
X8	Initial pH	6.5	7

X1-X8 = Medium constituent variable codes.

Table 3.3 shows the medium composition at its high and low design level for the culturing of *C. paraputrificum*.

Table 3.3 Levels of medium component variables for the culture of *C. paraputrificum* using Placket-Burman design.

Code	Variables	Low level (-)	High level (+)
X1	N-acetylglucosamine (g/l)	7.5	15
X2	Peptone (g/l)	3	4
X3	Yeast extract (g/l)	1	2
X4	Buffer capacity (M)	0.2	0.25
X5	L-cysteine.HCl (g/l)	0.5	1
X6	$FeSO_4.7H_2O$ (g/l)	0.1	0.4
X7	MgCl2 (g/l)	0.214	0.427
X8	Initial pH	6.5	7

X1-X8 = Medium constituent variable codes.

The C:N:P ratio for the initial medium concentration for *C. biejerinkii* culture was 66:9:20 in mmoles. The C:N:P ratio for the initial medium concentration for *C. paraputrificum* culture was 57:8:20 in mmoles. The Plackett and Burman design used in the research work is shown in table 3.4. There were 12 individual trials consisting of eight assigned variables (various constituents of the medium) and three unassigned variables, known as dummy variables. These dummy variables are used to calculate the experimental error when the data are analyzed. The dummy variables X_9, X_{10} and X_{11} were temperature at 37°C, inoculation volume and agitation at 120rpm.

Table 3.4 Plackett and Burman design for 12 trials.

Culture Trials	Medium constituent variables								Dummy variables		
	X_1	X_2	X_3	X_4	X_5	X_6	X_7	X_8	X_9	X_{10}	X_{11}
1	+	+	−	+	+	+	−	−	−	+	−
2	−	+	+	−	+	+	+	−	−	−	+
3	+	−	+	+	−	+	+	+	−	−	−
4	−	+	−	+	+	−	+	+	+	−	−
5	−	−	+	−	+	+	−	+	+	+	−
6	−	−	−	+	−	+	+	−	+	+	+
7	+	−	−	−	+	−	+	+	−	+	+
8	+	+	−	−	−	+	−	+	+	−	+
9	+	+	+	−	−	−	+	−	+	+	−
10	−	+	+	+	−	−	−	+	−	+	+
11	+	−	+	+	+	−	−	−	+	−	+
12	−	−	−	−	−	−	−	−	−	−	−

X_1 to X_8 = Medium constituent variables.
X_9 to X_{11} = Dummy variables.
Trials 1 – 12 = Experimental trial designs

The matrix is designed in a manner in which each independent variable is evaluated six times at its low (-) level and six times at its high (+) level. Similarly every time variable X_1 is evaluated with the high level value, variable X_2 is evaluated three times with its high value and three times with its low value. The same treatment is followed when variable X_1 is evaluated at its low value. This design therefore allows variable X_1 to be independently analyzed and evaluated as the matrix design cancels the effect of changing variable X_2 in the presence of variable X_1. Providing that there are   no obvious interactions between the independent variables, the nature of this design can be applied.

The Box-Behnken method was applied later in the work described in this work when the effects of interactions between the medium constituents needed to be studied in greater detail. The statistical software package Minitab version 16.1.0 (Minitab Inc, Pennsylvania, USA) was used to analyze the experimental results. Variance effects and standard errors were calculated and the Student's t-test was applied out to analyze the results. Analysis of variance was used to identify significant variables.

3.3 Results

3.3.1 HPLC analysis of standards

The qualitative and quantitative determination of putative metabolites can be undertaken by measuring the individual retention times and peak areas of these metabolites in a chromatogram, as demonstrated in Figure 3.1. From these data, the concentration of the metabolites in the unknown samples can be determined. The retention time of a solute, which is simply the time elapsed between the time of injection of a solute in a sample and the time of elution of the maximum peak height of that solute, is characteristic of each solute. Therefore each individual organic acid has a different retention time that can be used to identify it.

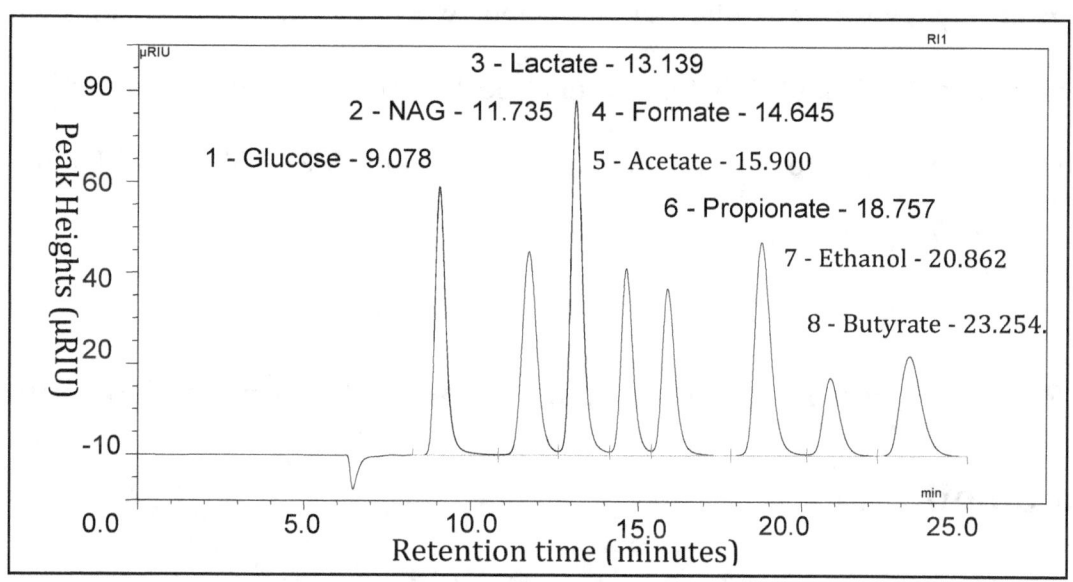

Figure 3.1 Standard chromatogram for HPLC analysis of the standard mixture of glucose, N-acetylglucosamine , lactate, formate, acetate, propionate, ethanol and butyrate.

Figure 3.1 shows a HPLC chromatogram of a mixture of standards for glucose (5 g/l), N-acetylglucosamine (5 g/l), lactate (10 g/l), formate (10 g/l), acetate (10 g/l), propionate (10 g/l), ethanol (5 g/l) and butyrate (5 g/l). Glucose is represented by the first peak in the chromatogram (fig. 3.1) with a peak area of 32.299 μRIU/min and was the first sugar to be eluted, with a retention time of 9.078 min. N-acetylglucosamine is represented by the second peak with a peak area of 31.883 μRIU/min and was the second sugar to be eluted, with a retention time of 11.735 min. The third peak represents lactate, which was the first organic acid to be eluted, had a retention time of 13.139 min and a peak area of 42.691 μRIU/min. Formate followed lactate with a retention time of 14.645 min and a peak area of 23.210 μRIU/min. Acetate was the 5th peak, and the third acid, followed by propionate with retention times of 15.900 min and 18.757 min, respectively, and peak areas of 30.442 μRIU/min and 36.673 μRIU/min, respectively. These were followed by ethanol, which eluted in 20.862 mins with peak area of 12.772 μRIU/min. The final peak represents butyrate, which was the last organic acid to elute and had a retention time of 23.254 min with a peak area of 20.402 μRIU/min.

3.3.2 Standard curves for N-acetylglucosamine and metabolites.

Figure 3.2 presents the calibration curves for the N-acetylglucosamine standards and also the metabolite (organic acid) standards. The calibration curves were performed in triplicate.

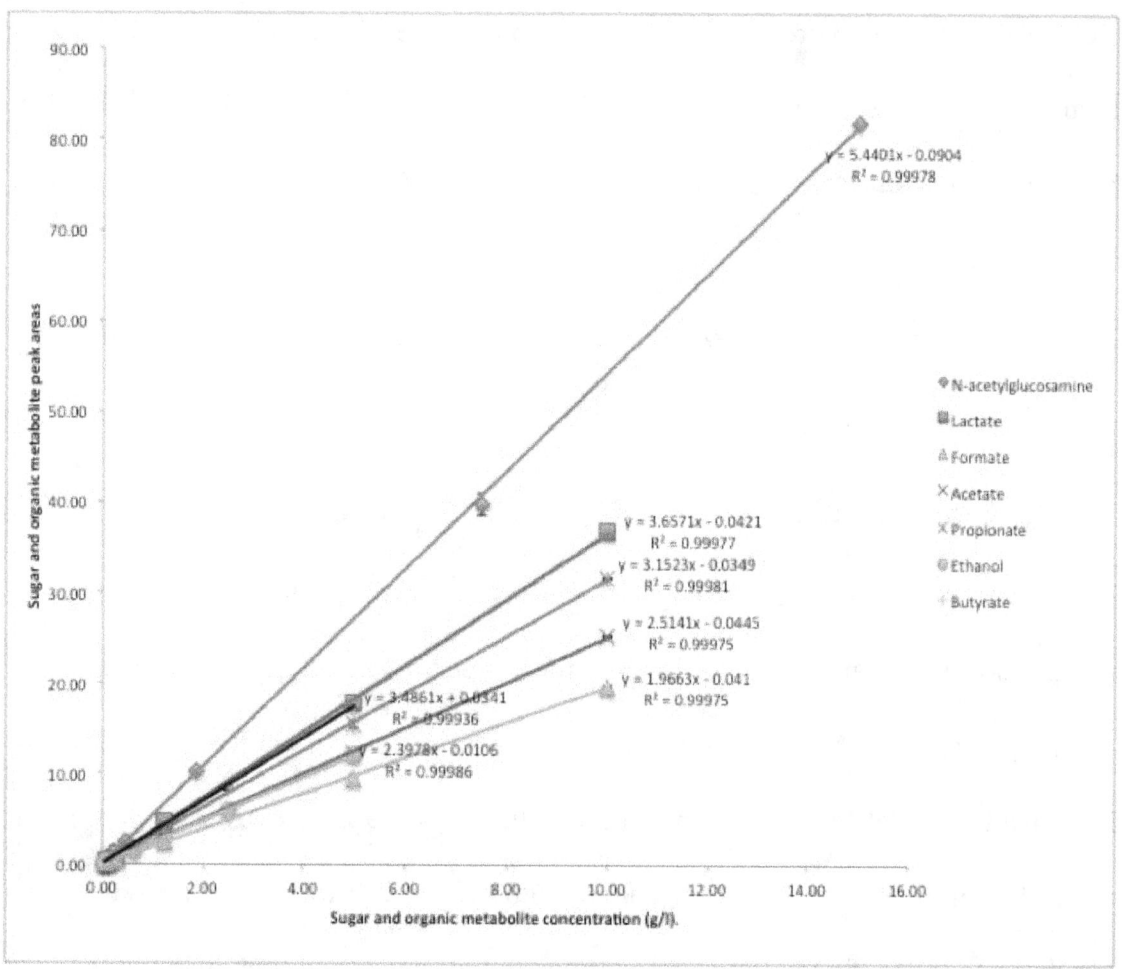

Figure 3.2 Calibration curves for N-acetyl glucosamine, lactate, formate, acetate, propionate, ethanol and butyrate standards derived from HPLC chromatograms.

n = 3
Conc (g/l) = Concentration of sugar and organic acids in grams per litre with standard errors

The peak areas obtained from running standards of N-acetylglucosamine, lactate, formate, acetate, propionate, ethanol and butyrate at 15 g/l, 10 g/l, 10 g/l, 10 g/l, 10 g/l, 5 g/l and 5 g/l, respectively, and at concentrations of 100%, 50%, 12.5%, 3.13%, 1.56%, 0.78% were used to plot standard curves of peak areas against concentration for N-acetylglucosamine and the metabolites. Figure 3.2 shows good linear correlations between the concentration of the standard metabolites and sugars and the area under the peak for each of the compounds tested

(Regression values were as follows: GlcNAc R^2 = 0.99978; lactate R^2 = 0.99977; formate R^2 = 0.99975; acetate R^2 = 0.99975; propionate R^2 = 0.99981; ethanol R^2 = 0.99986; R^2 = butyrate 0.99936).

3.3.3 Ability of *Clostridium beijerinckii* to utilise N-acetylglucosamine.

3.3.3.1 Six-day cultures showing carbon source utilization.

A total of 0.87 g/l of glucose had been used by the cultures of *Clostridium beijerinckii* by the end of the 6-day incubation period when grown in the glucose (5.07 g/l) positive control medium. At the same time the culture density had increased to 0.749AU from an initial value of 0.093AU with a concomitant reduction in the culture medium pH from 5.07 to 4.45 (Table 3.5). There was also an increase in the density of the negative control culture, which achieved a final density of 0.669 AU having started at 0.097 AU. There was only a slight reduction in pH from 5.10 initially to 4.71 at the end of fermentation (Table 3.5). The test cultures of *Clostridium beijerinckii* utilized 5.12 g/l of N-acetyl glucosamine. The density of the cultures grown in N-acetyl glucosamine increased from 0.086 AU, initially, to 1.659 AU after 6 days, and there was a reduction in the pH from 5.11 to 4.88 (Table 3.5). The cultures grown with GlcNAc showed a considerable difference from the glucose positive control cultures.

Table 3.5 *Clostridium beijerinckii* growth in the negative control, the glucose positive control, and N- acetylglucosamine 6 day culture.

Days	Negative control			Glucose			N-acetylglucosamine		
	O.D	pH	Conc (g/l)	O.D	pH	Conc (g/l)	O.D	pH	Conc (g/l)
0	0.097	5.10	0 ± 0.00	0.093	5.07	5.07 ± 0.07	0.086	5.11	5.13 ± 0.06
6	0.669	4.71	0 ± 0.00	0.749	4.45	4.20 ± 0.07	1.659	4.88	0.01 ± 0.00

O.D= Optical density (560 nm)
Conc (g/l) = Concentration of carbon source in grams per litre with standard errors.
 n = 3

3.3.3.2 Metabolic Product formation by *Clostridium beijerinckii* from utilizing N-acetylglucosamine.

Butyrate (3.20g/l) was the major organic acid produced by *C. beijerinckii* when supplied with N-acetylglucosamine as the sole sugar source, and this was followed by acetate (0.92g/l), ethanol (0.11g/l), lactate (0.08g/l) and propionate (0.05 g/l) (Table 3.5). Butyrate (0.96 g/l) was also the major organic acid produced when glucose was used as the sole sugar source. The organic acids produced in the negative controls were lower in concentration compared to the cultures grown with GlcNAc. Table 3.7 shows the increase in the culture optical density when N-acetylglucosamine was supplied as a substrate for *C. beijerinckii.*

Table 3.6 The concentration of organic acids produced by the *Clostridium beijerinckii* cultures in the negative control, glucose positive control and N-acetylglucosamine media.

Metabolic products (g/l)	Negative control	Glucose	N-acetylglucosamine
Lactate	0.00	0.00	0.08
Formate	0.04	0.00	0.00
Acetate	0.01	0.00	0.92
Proprionate	0.01	0.00	0.05
Ethanol	0.02	0.00	0.11
Butyrate	0.94	0.96	3.20

Conc (g/l) = Concentration of metabolic products in grams per litre.
n = 3

Table 3.7 *Clostridium beijerinckii* growth in the negative control, the glucose positive control, and N- acetylglucosamine in a 15 day culture.

Days	Negative control			Glucose			N-acetylglucosamine		
	O.D	pH	Conc (g/l)	O.D	pH	Conc (g/l)	O.D	pH	Conc (g/l)
0	0.033	6.03	0 ± 0.00	0.021	6.10	4.15 ± 0.13	0.024	6.12	4.41 ± 0.10
15	0.369	5.91	0 ± 0.00	2.352	5.15	1.70 ± 0.01	3.000	5.02	0.08 ± 0.01

O.D = Optical density (λ560)
Conc (g/l) = Concentration of carbon source in grams per litre with standard errors.
n = 3

A total of 2.45 g/l of glucose were utilised by the *C. beijerinckii* culture during 15-days of incubation, whilst in comparison 4.33 g/l of GlcNAc were utilised (Table 3.7). The *C. beijerinckii* biomass concentration increased from 0.021 AU to 2.352 AU when supplied with glucose, and from 0.024 AU to 3 AU when supplied with GlcNAc.

In addition, a total of 0.62 g/l for glucose and 5.12 g/l of GlcNAc were utilised in a 6-day culture in Table 3.5. The biomass concentration in absorbance unit for glucose increased from 0.093 AU to 0.749 AU and for GlcNAc was from 0.086 to 1.659 AU. Similar results were observed in table 3.5 and 3.7, as both showed that more GlcNAc was utilized compared with glucose utilized. There was also a greater increase in biomass concentration in the cultures grown with GlcNAc than those grown with glucose.

3.3.4 Plackett and Burman preparation.

It was observed, during preliminary experiments for the Plackett and Burman culture medium analysis that browning and precipitates occurred in some medium formulations during sterilisation. The constituent combinations that led to browning and precipitates in the medium formulations are noted in Table 3.8 below.

Table 3.8 Medium constituents that led to browning and precipitate formation after autoclaving of the medium formulation.

Medium constituents and autoclaving (g/l)	Results
Tryptone	No precipitate
Tryptone + L-cysteine.HCl	No precipitate
$MgSO_4$ + L-cysteine.HCl	No precipitate
$FeSO_4.7H_2O$ + Tryptone + L-cysteine.HCl	No precipitate
L-cysteine.HCl	No precipitate
$FeSO_4.7H_2O$ + Tryptone + L-cysteine.HCl + $MgSO_4$	No precipitate
$MgSO_4$ + Tryptone	No precipitate
$FeSO_4.7H_2O$ + L-cysteine.HCl	No precipitate
$FeSO_4.7H_2O$ + $MgSO_4$+ L-cysteine.HCl	No precipitate
$MgSO_4$ + L-cysteine.HCl + Tryptone	No precipitate
$MgSO_4$	Pale pink colour
$FeSO_4.7H_2O$	Light brown precipitate
$FeSO_4.7H_2O$ + $MgSO_4$	Dark brown precipitate
$FeSO_4.7H_2O$ + Tryptone + $MgSO_4$	Dark brown precipitate
$FeSO_4.7H_2O$ + Tryptone.	Dark brown precipitate

> Conc (g/l) = Concentration of carbon source in grams per litre with standard errors.

The presence of $FeSO_4.7H_2O$ in the medium led to both the browning and the formation of precipitates, whilst the presence L-cysteine.HCl.H_2O reduced this effect (Table 3.8). In medium formulations containing L-cysteine.HCl.H_2O there was no discolouration of the medium nor formation of precipitates. Therefore $FeSO_4.7H_2O$ was filter sterilized and added to the remainder of the medium, for the Plackett and Burman design experiments, after autoclaving.

3.3.5 Batch trial of medium designs for the growth of *Clostridium beijerinckii* and *C. paraputrificum* using Plackett and Burman design.

Different medium constituents and culture factors were investigated. The effect of different media constituent concentrations on product formation by *C. beijerinckii* and *C. paraputrificum* from N-acetylglucosamine as the carbon source in anaerobic batch culture were studied. These experiments were performed with initial GlcNAc concentrations of 10 g/l and 15 g/l for *C. beijerinckii* and 7.5 g/l and 15 g/l for *C. paraputrificum*, with the medium designed using Plackett and Burman's principles. The concentrations of organic acids produced into the medium by the Clostridium cultures were measured at the end of the anaerobic batch cultivation, when the formation of products and biogas had ended. Table 3.9 shows the result for comparing the experimental design with medium constituents at high and low values and the total biogas volume produced from different experimental trials.

Table 3.9 Plackett and Burman experimental design matrix for evaluating the factors influencing biogas production by *C. beijerinckii* and *C. paraputrificum*.

Culture Trials	Medium constituent variables								Biogas volume (ml/l)	
	X_1	X_2	X_3	X_4	X_5	X_6	X_7	X_8	*C. beijerinckii*	*C. paraputrificum*
1	+	+	−	+	+	+	−	−	1067	100
2	−	+	+	−	+	+	+	−	1850	200
3	+	−	+	+	−	+	+	+	0	1000
4	−	+	−	+	+	−	+	+	1433	167
5	−	−	+	−	+	+	−	+	1267	0
6	−	−	−	+	−	+	+	−	1917	633
7	+	−	−	−	+	−	+	+	2233	300
8	+	+	−	−	−	+	−	+	0	433
9	+	+	+	−	−	−	+	−	3133	1300
10	−	+	+	+	−	−	−	+	150	167
11	+	−	+	+	+	−	−	−	3183	0
12	−	−	−	−	−	−	−	−	1900	0

X_1 to X_8 = Medium constituent variables.
Trials 1 – 12 = Experimental trial designs
ml/l = biogas volume in millilitres per litre.
n = 3

Table 3.10 Profile of fermentation process parameters for *C. beijerinckii* after a 20-day incubation with N-acetylglucosamine as the main carbon source.

Trials	Biogas Volume (ml/l)	Biomass (g/l)	S.E	Initial pH	Final pH	S.E
T1	1067	0.35	0.03	6.5	5.94	0.19
T2	1850	0.26	0.05	6.5	6.19	0.01
T3	0	0.21	0.00	7.0	6.67	0.26
T4	1433	0.33	0.00	7.0	6.63	0.00
T5	1267	0.31	0.00	7.0	6.53	0.01
T6	1917	0.52	0.05	6.5	6.23	0.01
T7	2233	0.49	0.02	7.0	6.40	0.01
T8	0	0.23	0.00	7.0	6.87	0.01
T9	3133	0.79	0.05	6.5	5.96	0.01
T10	150	0.35	0.01	7.0	6.68	0.07
T11	3183	1.02	0.07	6.5	5.99	0.02
T12	1900	0.41	0.01	6.5	6.10	0.00

Trials 1 – 12 = Experimental trial designs
Biogas volume (ml/l) = Volume of gas produced per litre of medium.
S.E = Standard error
g/l = Biomass concentration in grams per litre.
n = 3

For cultures of *C. beijerinckii* the initial biomass was 0.163 ± 0.013 g/l. There was a decrease in pH in all of the experimental trials as well as an increase in biomass concentration. In trials T_3 and T_8 there was no biogas produced (Table 3.10) whilst only a small proportion of the GlcNAc available was utilized in these trials (Table 3.11). There were organic acids produced in all of the trials. These organic acids include lactate, acetate, ethanol and butyrate. Table 3.11 shows the concentration of organic acids produced compared with the GlcNAc utilized.

Table 3.11 Profile of N-acetylglucosamine utilization, and organic acid production by *C. beijerinckii* in trials T_1 to T_{12}.

Trial Number	Initial GlcNAc (g/l)	GlcNAc Utilised (g/l)	Lactate (g/l)	Acetate (g/l)	Ethanol (g/l)	Butyrate (g/l)	Biogas Volume (ml/l)
T1	15	14	1.40	4.71	0.00	3.70	1067
T2	10	10	0.00	2.99	0.24	3.38	1850
T3	15	4	1.48	0.64	0.00	0.00	0
T4	10	10	0.00	3.26	0.00	2.73	1433
T5	10	10	0.00	2.96	0.00	3.08	1267
T6	10	10	0.00	3.29	0.00	2.92	1917
T7	15	14	0.00	4.21	0.00	4.11	2233
T8	15	2	0.73	0.37	0.00	0.39	0
T9	15	15	0.00	3.91	0.00	4.99	3133
T10	10	5	1.63	1.43	0.00	1.24	150
T11	15	15	3.53	0.00	0.00	4.22	3183
T12	10	10	0.00	3.00	0.00	3.28	1900

Trials 1 – 12 = Experimental trial designs
GlcNAc = N-acetylglucosamine
g/l = concentration in grams per litre.
Biogas volume (ml/l) = Volume of gas produced per litre of medium.
n = 3

Acetate and butyrate were the most abundant organic acid produced (Table 3.11), followed by lactate and ethanol in all of the experimental trials. Trials T_9 and T_{11} produced the greatest volume of biogas and demonstrated the greatest GlcNAc utilization. There was a trend towards production of relatively large volumes of biogas from trial designs in which a large quantity of GlcNAc was found to be utilized. In contrast there tended to be little or no biogas production by cultures of *C. beijerinckii* from trials in which little of the sugar was utilized.

Table 3.12 Profile of fermentation process parameters for *C. paraputrificum* after a 20-day incubation with N-acetylglucosamine as the main carbon source.

Trials	Biogas Volume (ml/l)	Biomass (g/l)	S.E	Initial pH	Final pH	S.E
T1	100	0.11	0.01	6.5	5.74	0.03
T2	200	0.22	0.03	6.5	5.89	0.01
T3	1000	0.41	0.02	7.0	6.31	0.04
T4	167	0.18	0.01	7.0	6.59	0.01
T5	0	0.07	0.01	7.0	6.45	0.00
T6	633	0.33	0.01	6.5	6.10	0.00
T7	300	0.23	0.00	7.0	6.37	0.03
T8	433	0.28	0.02	7.0	6.44	0.02
T9	1300	0.55	0.03	6.5	5.57	0.03
T10	167	0.18	0.01	7.0	6.58	0.01
T11	0	0.07	0.00	6.5	6.53	0.01
T12	0	0.07	0.01	6.5	6.37	0.17

Trials 1 – 12 = Experimental trial designs
Biogas volume (ml/l) = Volume of gas produced per litre of medium.
S.E = Standard error
g/l = Biomass concentration in grams per litre.
n = 3

The initial biomass of *C. paraputrificum* was 0.057 ± 0.012 g/l. In addition, there was a pH declined from initial levels, whilst there was an increase in biomass concentration in all of the experimental trials (Table 3.12). Organic acids were also produced in all of the experimental trials; these organic acids included lactate, acetate, and butyrate, along with ethanol. Table 3.13 shows the concentration of organic acids produced compared with the GlcNAc utilized by *C. paraputrificum.*

Table 3.13 Profile of N-acetylglucosamine utilization, and organic acid production by *C. paraputrificum* in trials T_1 to T_{12}.

Trial Number	Initial GlcNAc (g/l)	GlcNAc Utilised (g/l)	Lactate (g/l)	Acetate (g/l)	Ethanol (g/l)	Butyrate (g/l)	Biogas Volume (ml/l)
T1	15.0	9.4	3.75	3.07	0.00	0.51	100
T2	7.5	7.5	1.90	2.70	0.00	0.91	200
T3	15.0	13.9	2.83	4.74	0.00	1.75	1000
T4	7.5	7.5	1.77	2.53	0.00	0.86	167
T5	7.5	7.5	3.39	2.43	0.00	0.00	0
T6	7.5	7.5	1.38	2.67	0.00	1.47	633
T7	15.0	10.4	2.28	3.57	0.00	1.53	300
T8	15.0	10.3	2.21	3.41	0.00	1.34	433
T9	15.0	13.1	2.79	4.40	0.71	1.73	1300
T10	7.5	7.4	2.38	2.43	0.51	0.56	167
T11	15.0	0.4	0.00	0.00	0.00	0.00	0
T12	7.5	0.0	0.00	0.00	0.00	0.00	0

Trials 1 – 12 = Experimental trial designs
GlcNAc = N-acetylglucosamine
g/l = concentration in grams per litre.
Biogas volume (ml/l) = Volume of biogas produced per litre of medium.
n = 3

During the cultivation of *C. paraputrificum* ethanol was detected as a minor by-product at low concentrations compared with the other metabolic products. Acetate was the most abundant organic acid produced, followed by lactate and then butyrate. Trials T_3 and T_9 produced the highest volume of biogas and also utilized the most GlcNAc. In experimental trials T_5, T_{11} and T_{12} no biogas was produced and GlcNAc was not utilized (Table 3.13). Trials with low sugar utilization produced a low quantity of biogas or no biogas; this was the same for both *C. beijerinckii* and *C. paraputrificum*. Table 3.14 and 3.15 show the levels of the variables and statistical analysis of the Plackett and Burman design. The statistical analysis were used to identify the significant medium variables for the Plackett and Burman design for *C. beijerinckii* and *C. paraputrificum* cultures.

Table 3.14 Statistical analysis of the results from the Plackett and Burman medium design experiments for *C. beijerinckii*, along with the variables tested.

Code	Variables	Low level (-)	High level (+)	Effect (Exi)	t-values	Prob>|t|
X1	N-acetylglucosamine (g/l)	10	15	175.0	0.45	0.683
X2	Peptone (g/l)	3	4	-486.0	-1.25	0.300
X3	Yeast extract (g/l)	1	2	172.0	0.44	0.690
X4	Buffer capacity (M)	0.2	0.25	-439.0	-1.14	0.337
X5	L-cysteine.HCl (g/l)	0.5	1	655.0	1.70	0.188
X6	$FeSO_4.7H_2O$ (g/l)	0.1	0.4	-989.0	-2.56	0.083
X7	$MgCl_2$ (g/l)	0.214	0.427	499.0	1.29	0.288
X8	Initial pH	6.5	7	-132.7	-3.44	0.041

X_1 to X_8 = Medium constituent code
(-) = Medium constituent at a low level
(+) = Medium constituent at a high level
Exi = Effect of a constituent variable.
t-values = T test values for each variable.
Prob > |t| = Probability of t
g/l = concentration in grams per litre.
n = 3

Table 3.15 Statistical analysis of the results from the Plackett and Burman medium design experiments for *C. paraputrificum*, along with the variables tested.

Code	Variables	Low level (-)	High level (+)	Effect (Exi)	t-values	Prob>|t|
X1	N-acetylglucosamine (g/l)	7.5	15	328.0	2.56	0.083
X2	Peptone (g/l)	3	4	72.0	0.56	0.613
X3	Yeast extract (g/l)	1	2	172.3	1.35	0.271
X4	Buffer capacity (M)	0.2	0.25	-28.0	-0.22	0.841
X5	L-cysteine.HCl (g/l)	0.5	1	-461.0	-3.60	0.037
X6	$FeSO_4.7H_2O$ (g/l)	0.1	0.4	72.3	0.56	0.612
X7	$MgCl_2$ (g/l)	0.214	0.427	483.3	3.77	0.033
X8	Initial pH	6.5	7	-28.0	-0.22	0.841

X_1 to X_8 = Medium constituent code
(-) = Medium constituent at a low level
(+) = Medium constituent at a high level
Exi = Effect of a constituent variable.
t-values = T test values for each variable.
Prob > |t| = Probability of t.
g/l = concentration in grams per litre.
n=3

The effect of a variable on a response was determined by the difference between

the average of the total of high (+)   responses from the average of the total of low (-) responses (Greasham & Herber, 1997). Effects of the dummy variables were calculated to be 6.21, 2.32, and 0.996 (*C. beijerinckii*) and -0.280, 1.276, and 1.166 (*C. paraputrificum*) for X_9, X_{10} and X_{11} respectively. The variance effect was calculated by averaging the squares of the dummy variables effects and this was calculated to be 14.97 (*C. beijerinckii*) and 0.969 (*C. paraputrificum*). Standard error was calculated as the square root of the variance and was determined to be 3.86 (*C. beijerinckii*) and 0.9848 (*C. paraputrificum*).

$$SE = \sqrt{V_{eff}} \quad \text{.................. Equation 3.1}$$

A Student's T test was carried out for each variable and this was determined by dividing its individual effect by the standard error (SE) as shown in the equation below:

$$tx_1 = Ex_1/SE. \quad \text{.....................Equation 3.2}$$

The results for the t-values are shown in table 3.14 and 3.15. The importances of the variables, which are the medium components assigned high and low values were investigated using the Plackett and Burman method. The response (total biogas volume) produced from each trial is shown in Tables 3.11 and 3.13. The extent of a variable effect on a response is also shown and is denoted by a positive or a negative sign (Table 3.14 and 3.15).

For *C. beijerinckii* (Table 3.14) it was shown that high levels of X_1 (N-acetyl glucosamine), X_3 (yeast extract), X_5 (L-cysteine.HCl.H_2O) and X_7 (Magnesium chloride) enhanced the production of biogas and low levels of X_2 (peptone), X_4 (buffer capacity), X_6 ($FeSO_4.7H_2O$) and X_8 (initial pH) also resulted in higher biogas volume. Whilst for *C. paraputrificum* (Table 3.15) it was shown that high levels of X_1 (N-acetylglucosamine), X_2 (peptone), X_3 (yeast extract), X6 (FeSO$_4$.7H$_2$0) and X_7 (Magnesium chloride) enhanced the production of biogas and at low levels of X_4 (Buffer capacity), X_5 (L-cysteine.H2O) and X_8 (initial pH) also enhanced the production biogas. Consulting the T-distribution table (two sided), with 3 degrees of freedom, medium component variables X_1, X_2, X_3, X_4, X_5 and X_7 (*C. beijerinckii*) in table 3.14 and component variables X_2, X_3, X_4, X_6 and X_8 (*C. paraputrificum*) in table 3.15 were considered to be insignificant since they demonstrated confidence levels below 90%. Variables X_6 and X_8 (*C. beijerinckii*) and X_1, X_5 and X_7 (*C. paraputrificum*) with confidence levels above 90% were considered significant. Variables with significant effects were considered in the next stage of medium optimization.

The Plackett and Burman design experiments permitted the significant constituents of the growth medium to be identified, and so a suitable medium composition was developed for both *C. beijerinckii* and for *C. paraputrificum*, as described in Table 3.16.

It should be noted that these are not the fully optimised levels of the individual factors involved; further and more intensive optimization steps need to be carried out. Only those significant medium constituents and the concentrations of those that gave the greatest effects were identified. Further optimization is required to locate the vicinity of optimum response, which will be done using the Method of Steepest Ascent, and then the Box-Behnken method will be used to identify the optimum medium concentrations that give the highest total volume of biogas. In addition the interaction effects between these medium constituent variables will be assessed.

Table 3.16 The proposed medium compositions for the growth of *C. beijerinckii* and *C. paraputrificum* based on the results from the Plackett and Burman design experiments.

Variable Code	Medium Variables	*C. beijerinckii*	*C. paraputrificum*
		Variable concentrations	
X1	N-acetylglucosamine	15 g/l	15 g/l
X2	Peptone	3 g/l	4 g/l
X3	Yeast extract	2 g/l	2 g/l
X4	Buffer capacity	0.2 M	0.2 M
X5	L-cysteine.HCl.H$_2$O	1 g/l	0.5 g/l
X6	FeSO$_4$.7H$_2$0	0.1 g/l	0.4 g/l
X7	MgCl$_2$	0.43 g/l	0.43 g/l
X8	Initial pH	6.5	6.5

g/l = concentration in grams per litre.
X1 – X2 = Medium variable code.

3.4 Discussion.

3.4.1 Ability of *C. beijerinckii* to utilize N-acetylglucosamine.

Over a six-day incubation period *C. beijerinckii* was able to utilize N-acetylglucosamine and over that incubation period, the cell density substantially increased. During this incubation period *C. beijerinckii* also produced organic acids from the metabolism of GlcNAc. Butyrate was the major organic acid produced followed by acetate, ethanol, lactate and propionate when N-acetyl glucosamine was utilized (Table 3.6). Butyrate was also the major organic acid produced when glucose was utilized as the sole sugar source. Xin et al. (2011) investigated hydrogen production by the newly isolated *Clostridium beijerinckii* RZF-1108, and the main organic acids produced from glucose

110

fermentation were acetic acid, butyric acid and lower concentrations of ethanol and butanol. They also suggested that the organic acids produced were responsible for the decline in pH. Simunek, Tishchunko and Koppova (2008) showed that *C. beijerinckii* is able to utilize colloidal chitin and produces a variety of chitinolytic enzymes. Consequently, it is not surprising that it has the ability to metabolize N-acetyl glucosamine, although to this author's knowledge this is the first time that this has been recorded.

The results of the experiments described here indicate that *C. beijerinckii* can utilise N-acetylglucosamine as its sole carbon source for metabolism and cell multiplication. As expected the pH of the *C. beijerinckii* negative control culture grown without a carbon source, did not decrease substantially over the course of the experiment. There was a greater final cell density, and the production of much higher levels of organic acids in the N-acetylglucosamine supplied cultures, than in the negative control cultures.

3.4.2 Medium constituents causing precipitates to develop.

From the results it was concluded that $FeSO_4.7H_2O$ caused medium precipitation due to the brown coloration. Bartram and Balance (1996) stated the precipitation caused by iron is due to oxidation from the soluble ferrous state (+2) to the ferric state (+3) that precipitates as insoluble hydrated ferric oxide. Once iron is exposed to oxygenated environmental conditions in water, Fe^{2+} is readily oxidized by the actions of molecular oxygen as described in equation 3.3.

$$4Fe^{2+} + O_2 + 4H^+ -> 4Fe^{3+} + 2H_2O. .... Equation\ 3.3$$

Consequently hydrolysis occurs, which then produces crystalline Fe (III) oxides, which have low solubility and low solubility product constant (Ksp) values (Tuhela et al., 1993). The Ksp value indicates the degree to which a compound dissociates in water. The higher the solubility product constant, the more soluble it is. In contrast, when L-cysteine.HCl.H$_2$0 was present in the medium the development of a brown colour and the production of precipitates was reduced.

3.4.3 Preparation of inoculum for Placket and Burman method.

It is known that the volume of the initial inoculum culture has an effect on the length of the lag phase. The initial inoculum volume was increased to increase the cell density in the inoculum and thereby to reduce the lag phase. Experiments with *C. beijerinckii* and *C. paraputrificum* were performed with a relatively large initial inoculum to minimize the lag phase. In the present work these procedures were used with additional steps where the inoculum from starter cultures were centrifuged and then resuspended in the specific Plackett and Burman trial media. Washing and resuspension of the initial inoculums in the specific medium would help to reduce or remove organic acids, excreted microbial compounds and unwanted contaminant medium constituents from the inoculum. Preparation of the initial inoculums in this way would help to make the results of the Plackett and Burman method more accurate, although it may also increase the lag phase demonstrated by *C. beijerinckii* and *C. paraputificum.*

3.4.4 Plackett and Burman Design for *Clostridium beijerinckii* medium.

The comparative importance of N-acetylglucosamine, peptone, yeast extract, $FeSO_4.7H_2O$, buffer capacity, initial pH, and L-cysteine.$HCl.H_2O$ concentration for biogas production were investigated using the Plackett and Burman design. The pH reduction from all experimental trials (Table 3.10) was due to the formation of organic acids produced from the metabolism of N-acetylglucosamine. *Clostridium beijerinckii* when grown in medium design trials T_3 and T_8 was unable to utilize N-acetylglucosamine effectively (Table 3.12), compared with the other medium design trials, which reflected on a lack of biogas production. Similar medium components for T_3 and T_8 were high variable levels of N-acetylglucosamine, $FeSO_4.7H_2O$, Initial pH and low variable levels of L-cysteine.$HCl.H_2O$ (Table 3.9). It is known that *C. beijerinckii* produces hydrogen gas during the exponential phase of growth as the rate of carbon substrate utilization increases. Xin et al. (2011), furthermore, have reported that hydrogen production corresponds to *Clostridium beijerinckii* cell biomass yield. Trial T_{10} produced a small volume of biogas; this was also as a result of the low quantity of N-acetylglucosamine consumed in this medium design (Table 3.12). These results show that medium design trials T_3, T_8 and T_{10} were not favourable to high total volumes of biogas produced by *C. beijerinckii*. The limited quantities of

113

biogas produced in medium design trials T_3, T_8 and T_{10} correlate with the low final cell densities achieved by *C. beijerinckii*, as was observed previously by Xin et al. (2011).

In contrast, large volumes of biogas were produced by *C. beijerinckii* during medium design trials T_9 and T_{11}, and this correlated with the larger quantities of GlcNAc utilized and also larger final cell density achieved by the cultures in these experiments (Table 3.13). If the sign of the effect of a variable on a response (EXi) is positive, it means the influence of the variable on total biogas produced is greater at a high constituent level and if the sign is negative, the influence of the variable is greater at a low constituent level (Table 3.14). In other words when variables X_1 (N-acetylglucosamine), X_3 (yeast extract), X_5 (L-cysteine.HCl.H_2O) and X_7 ($MgCl_2$) are used at high levels, it means that these variables lead to greater biogas production than when they are used at low levels. On the other hand, when variables X_2 (peptone), X_4 (buffering capacity), X_6 ($FeSO_4.7H_2O$) and X_8 (initial pH), are used at low levels they lead to the production of greater total volumes of biogas than when they are used at high levels (Table 3.14). Greasham & Herber (1997) reported that an effect is considered real when the effect is significant, meaning the effect of the variable involved did not occur by chance but occurred as a result of the nutrient involved. Consequently, since X_6 ($FeSO_4.7H_2O$) and X_8 (initial pH) both led to a significant increase in biogas production, was as a result of the variable involved in the fermentation process and did not occur by chance. The significant effect is considered real (Greasham & Herber, 1997).

3.4.4.1 Effect of initial pH on total biogas volume.

Medium design trials T_9 and T_{11}, during which the greatest total volumes of biogas were produced, had an initial pH of 6.5. In contrast, medium design trials T_3, T_8 and T_{10}, during which little or no biogas was produced, started from an initial pH of 7. Generally, the pH of the growth medium will affect growth and so can influence the biosynthesis of metabolites.

Greasham & Herber (1997) reported that the optimum pH range for most bacteria growth is from 6.5 to 7.5. Pan et al. (2008) reported that for *C. beijerinckii* Fanp3 the optimum pH range for hydrogen production was from 6.47 to 6.98; this pH range led to production of 246.2–252.9 ml H_2/g-glucose. Pan et al. (2008) also stated that the initial pH could influence the lag time significantly, with a prolonged lag phase occurring when the initial pH was sub-optimal. The shortest lag time was obtained at initial pH 6.47.

From the Plackett and Burman result medium design trials, the shortest lag time before growth was observed occurred in trial T_9, which had the shortest lag time at an initial pH of 6.5 with and led to the production of a high large volume of biogas produced. The results show that there was a substantial influence by the initial pH on biogas production, as the effect of the variable was considered significant at a p value of 0.041. This simply means the effect of the variable involved did not occur by chance. Therefore, the effect caused by X_8 (initial pH), was as a result of the variable involved in the fermentation process. Dabrock et al. (1992) stated that pH may directly affect the activity of the hydrogenase enzyme required for hydrogen production.

Cai et al. (2010) performed a metabolic flux analysis to determine the effect of pH and glucose concentration on hydrogen production by *C. butyricum* W5. It was confirmed that *C. butyricum* W5 grew best at pH of 6.5 (Cai et al., 2010). In addition, the metabolite production profile was compared at pH 6.5 and pH 7.0, and it was found that all of the metabolites were produced much more slowly at pH 7 than at pH 6.5. This indicates that the initial pH significantly affected the biogas production because the pH may directly affect hydrogenase activity (Dabrock et al. 1992) and no trials in this study resulted in a lack of or low volume of final biogas when pH was set at an initial pH of 6.5 in comparison to the result from initial pH set at 7.

3.4.4.2 Effect of Iron (II) sulphate heptahydrate ($FeSO_4.7H_2O$) on biogas production.

During initial attempts to prepare the various growth medium, it was noted that the presence of iron caused a precipitate to develop in the medium. Precipitates were not observed in the medium trial designs, however trials with higher levels of $FeSO_4.7H_2O$ demonstrated a darker colouration of the medium. The effect of $FeSO_4.7H_2O$ was considered significant at a p-value 0.083. In the results of the Plackett and Burman method the effect of Fe^{2+} was significant at a low level of 20 mg/l (0.1 g/l $FeSO_4.7H_2O$). Generally Fe^{2+} is required by bacterial cells in trace quantities, but excess levels of Fe^{2+} (in the form of $FeSO_4.7H_2O$) in a medium may lead to precipitation of medium constituents. Precipitation of medium constituents out of solution would lead to them being unavailable for microbial nutrition and could lead to spurious results from the Plackett and Burman trials. The effect caused by X_6 (FeSO$_4$.7H$_2$O), was as a result of the variable concentration involved in the fermentation process. It has been suggested that environmental Fe^{2+} concentrations may not only affect hydrogen production directly, but also affect hydrogen production through other environmental perturbations (Lee et al., 2001). Investigations of the effect of environmental parameters on hydrogen production using *Clostridium saccharoperbutylacetonicum* N1-4 (ATCC 13564) demonstrated that iron-sulfur influenced protein functions primarily by acting as an electron carrier and was also involved in the oxidation of pyruvate to acetyl-CoA, CO_2 and H_2 (Alayah et al., 2009). The study of Alayah et al. (2009) concluded that the best optimum culture conditions for greatest hydrogen yield were: an initial of pH 6.0; an incubation temperature of 37°C; and the addition of ferrous sulfate to give 25 mg/l Fe^{2+}.

3.4.4.3 Effect of N-acetyl glucosamine concentration on biogas production

N-acetylglucosamine was utilized by *C. beijerinckii*, for the production of cell mass, metabolites and energy. Medium design trials T_9 and T_{11} demonstrated the greatest biogas production and also the greatest GlcNAc utilization (Table 3.11). It could be suggested that the medium trial designs were favorable for T_9 and T_{11} cultures to utilize GlcNAc for energy, thereby encouraging increase in cell density and biogas production. On the other hand cultures in T_3 and T_8 did not utilize a large quantity of GlcNAc, resulting in low biogas production. Zhao et al. (2011) reported that the hydrogen yield and production rate in *C. beijerinckii* RZF-1108 fermentation increased with increasing glucose supplementation from 5 g/l to 9 g/l and reached a maximum of 1.80 mol H_2/mol glucose. The Plackett and Burman medium design experiments showed that the effect of X_1 (GlcNAc) at a high level (15 g/l) was greater than at the low level (10 g/l) but this effect was not statistically significant because the confidence level was below 90%. This might also be as a result of the design of the high and low level differential as Greasham & Herber (1997) stated that a high differential is always recommended but if it is too large the effect of the variable might mask the effect of other variables. Therefore it may be safe to suggest that if the differential is too small, it may lead to an inability to determine the effects of high or low variables. Complementing the results from the Plackett and Burman to prove that the effect of N-acetylglucosamine should not be neglected, as its effect was greater at a high level; the ranges of effect were obtained. From 0 to 3183 ml of biogas per litre of culture medium were produced when the N-acetylglucosamine was at the high level, whilst for the low level a range of from 150 to 1917 ml of biogas per litre of culture medium was produced. This suggests some other, as yet uncharacterized, interactions were having an effect and masking the true impact of N-acetylglucosamine, since difference between a final biogas level of 3183 ml/l and one of 1917 ml/l is substantial even if not statistically significant in this case. Consequently, the next phase of optimization experiments was designed to take this possibility into account.

For the *C. beijerinckii* cultures in the Plackett and Burman medium design trials it would seem that L-cysteine.HCl.H$_2$O had the third greatest effect. The role of L-cysteine.HCl.H$_2$O is to reduce the dissolved oxygen effect in the culture medium and to help maintain anaerobic conditions. This suggests that oxygen ingress may be an issue in the experimental procedures. Therefore it was decided to investigate the effect of L-cysteine.HCl.H$_2$O on product formation, and to determine the optimum concentration required to keep the batch system anaerobic for *C. beijerinckii*, by using various L-cysteine.HCl.H$_2$O concentrations in batch culture experiments.

3.4.5 Plackett and Burman design for *Clostridium paraputrificum* medium.

The relative importance of N-acetylglucosamine, peptone, yeast extract, FeSO$_4$.7H$_2$0, buffer capacity, initial pH, and L-cysteine concentration on biogas production by *C. paraputrificum* were studied using the Plackett and Burman design. The formation of organic acids during metabolism of N-acetylglucosamine in all experimental trials led to the decline of the pH of the culture from its initial level (Table 3.11). *Clostridium paraputrificum* failed to utilize N-acetylglucosamine completely in medium design trials T$_5$, T$_{11}$ and T$_{12}$, and there was no biogas production. It is known that during the exponential phase of *C. paraputrificum* growth, the hydrogen gas production rate increases as its rate of carbon substrate utilization increases. For example, it has been reported that *C. paraputrificum* grew rapidly, doubled in biomass every 30 minutes and produced hydrogen gas when GlcNAc was the sole carbon source Evvyernie et al. (2000). Furthermore, hydrogen production by

Clostridium paraputrificum has been found to correlate with cell biomass yield (Xin et al., 2011). Some of the trial runs in the work described in this work produced only a small volume of biogas; this might have resulted from limited consumption of the N-acetylglucosamine. Consequently, these medium designs are unlikely to favour high yields of biogas production. In addition, only limited quantities of biogas were produced in some of the medium design trial cultures (T_1, T_5, T_{11}, T_{12}), which were correlated with low final cell densities.

On the other hand, medium design trials T_3 and T_9 resulted in the production of a large volume of biogas by *C. paraputrificum*, and this was correlated with utilization of a large proportion of the GlcNAc available. In addition, there was a relative high cell density at the end of the fermentation process in trials T_3 and T_9 (Table 3.12). The positive sign on the effect of a variable means the influence of the variable on total biogas produced is greater at a high concentration and if the sign is negative, the influence of the variable is greater at a low concentration (Table 3.15). The influence of variables X_1 (N-acetylglucosamine), X_2 (peptone), X_3 (yeast extract), X_6 (FeSO4.7H20) and X_7 (MgCl$_2$), when used at the high level (+ve), was greater than when they were used at the low level (-ve), because the sign of the effect (Exi) was positive; Table 3.15. In contrast, variables X_4 (Buffer capacity), X_5 (L-cysteine.HCl.H$_2$O) and X_8 (initial pH), produced a greater effect when used at the low level than at the high level. This simply means more volume of biogas was produced when these variables are used at the high or low values which the negative or positive (effect Exi) signs relates to in the Plackett and Burman analysis shown in table 3.15. An effect is considered real when the effect is significant, meaning the effect of the variable involved did not occur by chance but occurred as a result of the nutrient involved (Greasham & Herber, 1997). Therefore, the effects caused by X_1 (N-acetylglucosamine), X_5 (L-cysteine.HCl.H_2O) and X_7 (MgCl$_2$), were as a result of these variables involved in the fermentation process and did not occur by chance. The effects of significant variables are explained in the following sections.

3.4.5.1 The effect of N-acetylglucosamine concentration on biogas production by *C. paraputrificum.*

N-acetylglucosamine was utilized as a carbon source by *C. paraputrificum* for the production of cell mass, metabolites and energy. Generally, during the process of metabolizing a carbon source, high-energy phosphate compounds such as ATP are formed, for example by substrate level phosphorylation (Greasham & Herber, 1997). These compounds are consumed for cellular biosynthesis and cell maintenance. The concentration of a carbon source is also vital to successful medium design to encourage production of the desired metabolites.

The *C. paraputrificum* grown in medium design trials T_3 and T_9 demonstrated the greatest biogas yield in addition to the greatest GlcNAc utilization (Table 3.13). It appeared that the medium trial design favored cultures to utilize GlcNAc for energy, thus resulting in increased cell density and total biogas production. On the other hand the *C. paraputrificum* cultures grown in trials T_5, T_{11} and T_{12} failed to metabolise GlcNAc and they also yielded little biogas.

Evvyernie et al. (2001) have previously reported that *C.*

paraputrificum was able to utilize N-acetylglucosamine and recorded that the hydrogen production rate declined as a result of exhaustion of the limiting substrate (N-acetylglucosamine). This indicates that the exhaustion or the inability to utilize N-acetylglucosamine may lead to little or no biogas produced. The results from the Plackett and Burman experimental design trials described in this work showed that the effect of GlcNAc (X_1) on total biogas volume was significantly greater when it was supplied at the high level (15g/l) than when it was supplied at the low level (P = 0.083).

3.4.5.2 Effect of L-cysteine.HCl.H$_2$O on biogas production by *C. paraputrificum*.

Generally, obligate anaerobes will suffer or die when exposed to oxygen; this is because they lack the enzymes catalase and superoxide dismutase which helps to convert harmful superoxide formed in cells to oxygen when they are exposed to air. Clostridia are obligate anaerobes, which are not able to initiate growth in culture except when the medium is controlled to an appropriately low redox potential. Consequently, it is necessary to degas the medium and to eliminate the dissolved oxygen to avoid its toxicity, however degassing alone cannot achieve suitably low redox potentials for clostridial growth. Hence a reducing agent, such as cysteine, is normally included in the culture medium for clostridia and other strict anaerobes (Koesnandar, Nishio and Nagai, 1990). Variable X_5 (L-cysteine.HCl.H$_2$O) with confidence level above 90% was considered significant. Inclusion of L-cysteine.HCl.H$_2$O in the medium seemed to be vital and important in the medium design at a low concentration. Cysteine has been reported to stimulate the growth and fermentation rate of *Bacteroides ruminicola* GA33, *Bacteroides ruminicola* 23, *Bacteroides fibrisolvens* D1, *Eubacterium ruminantium* GA195, and *Streptococcus bovis* Pe1 (Jones et al., 1980). When *C. beijerinckii* was cultured in the high level of cysteine, growth of the culture was observed sooner than when it was cultured with the low level of cysteine (Figure 4.1). However those cultures which were cultivated with the high cysteine level produced a lower total volume of biogas and also a lower biomass than those cultivated with the low level of cysteine. This is reflected in the significant effect of L-cysteine.H_2O at the low level which contrasts with the lack of a significant effect at the high level.

Various reducing agents have been included as medium constituents to control the redox potential in media used to culture obligate anaerobes. For example cysteine hydrochloride has been used most extensively either alone or in combination with hydrogen sulphide or sodium sulphide. Exposure to atmospheric oxygen of a medium for anaerobes that contains cysteine, though, might result in a bactericidal effect because of hydrogen peroxide formation (Jones et al., 1980). Jones et al. (1980) also noted that when the medium contained no reducing agent, the maximum optical density achieved by bacterial cultures was often lower than, or only achieved after a long lag phase, compared with when cysteine hydrochloride was included in the medium constituents, which corresponds to the results achieved with cysteine in this work (Figure 4.1). In the current work, a reduction in fermentation lag times was achieved with increased L-cysteine.$HCl.H_2O$ concentrations, suggesting that L-cysteine.HCl.H$_2$O permitted cell growth to be initiated sooner by making the culture medium anaerobic. This may be because the L-cysteine.HCl.H$_2$O reduced the redox potential of the growth medium to a favorable level for biogas production (Sim & Kamaruddin, 2008). L-cysteine.HCl.H$_2$O was identified as a significant component of the medium, which aided by stimulating growth of the *C. paraputrificum* in the Placket and Burman medium design trials. It is likely that cysteine reduced the redox potential of the cultivation medium and so made the culture conditions more favourable for *C. paraputrificum* growth.

3.4.5.3 Effect of Magnesium Chloride.

Generally, during fermentation projects, the main focus of the work is to improve the yield of fermentation products, whilst one of the most neglected areas of complete fermentation medium design is the nutritional components needed to maintain healthy and viable cells. MgCl$_2$, at the high level (0.427 g/l), was found to have a significant effect upon the total volume of biogas produced in the Plackett and Burman medium design trials (P = 0.033). Generally MgCl$_2$ is required in trace quantities. In experiments performed to study the effect of various metal ions on *Clostridium perfringens* in different stages of its life-cycle - the vegetative growth stage and spore stage - it was found that Mg^{2+} is needed for the growth of vegetative *Clostridium perfringens* cells as well as for maintaining normal cellular morphology (Zhonghua et al., 1978).

Xin et al. (2012) studied the effects of Fe^{2+} and Mg^{2+} ions, and also L-cysteine, on hydrogen production and cell growth of *C. beijerinckii* RZF-1108 and also the expression of the Fe-hydrogenase (*hydA*) gene. Their work determined that inclusion of Mg^{2+} in the growth medium was responsible for increased bacterial cell growth and hydrogen production. However, the rate of hydrogen production, and expression of the *hydA* gene, were reduced by the presence of Mg^{2+}; the increased volume of hydrogen produced was because of an increase in cell biomass (Xin et al., 2012). The initiation of bio hydrogen production system and effects caused by the inclusion of metal Ions in culture medium for hydrogen production were also investigated, Fe^{2+} and Mg^{2+} were able to enhance the hydrogen production capacity of micro flora (Anying et al., 2010). Taking into account the evidence from the current study, and the literature quoted above, it would seem that it is important to include Mg^{2+} at a suitable concentration in a medium formulation to achieve the desired response.

3.4.6 Conclusion

Clostridium beijerinckii proved to be able to metabolize N-acetylglucosamine. The statistical methodology used for biogas production was the Plackett and Burman method to screen for significant medium constituent variables. The experimental results showed that the $FeSO_4.7H_2O$ concentration and the initial pH of the medium were significant factors for biogas production by *C. beijerinckii*, whilst for *C. paraputrificum* it was the N-actylglucosamine, L-cysteine.$HCl.H_2O$ and $MgCl_2$ concentrations that were significant factors.

Clostridium beijerinckii and *C. paraputrificum* both produced organic acids. Butyrate, acetate, lactate and ethanol were detected as by-products.

125

Generally butyrate and acetate were the most abundant organic acids produced when N-acetylglucosamine was the sole sugar source supplied to the cells in the Plackett and Burman design.

Chapter 4

One Factor L-Cysteine.HCl.H$_2$0 Medium Optimisation Design

4.1.1 Background

In the experiments described in this chapter the effect of different concentrations of L-cysteine.HCl.H$_2$O on total biogas produced were investigated. In the Plackett and Burman medium design experiments for *Clostridium beijerinckii,* it became apparent that L-cysteine.HCl.H$_2$O was the variable with the third greatest effect on total volume of biogas produced. However the role of L-cysteine.HCl.H$_2$O is to act as a reducing agent in the medium, effectively making it anaerobic. This suggests that oxygen ingress may be an issue in the experimental procedures. Therefore to investigate the formation of products (biogas and organic acids) and to determine the optimum concentration of L-cysteine.HCl.H$_2$O required to support maximal total volume of biogas of *C. beijerinckii*, the cultures were incubated with various L-cysteine.HCl.H$_2$O concentrations in the growth medium. The approach followed in this chapter will only focus on the effect of L-cysteine.HCl.H$_2$O concentration and will not investigate the interactions of other medium components.

The one factor technique was used by Wang, et al. (2008) to analyze the effect of different Fe^{2+} concentrations on mixed cultures using glucose as the main source of carbon. The range of Fe^{2+} concentrations used was from 0 to 1500 mg/l for fermentative hydrogen production in batches. At certain Fe^{2+} concentrations within the range, the hydrogen volume accumulated (302.3 ml) and the hydrogen yield (311.2 ml/g glucose) were much higher than at other concentrations. The maximum hydrogen volume accumulated, and the maximum hydrogen production yield, were achieved at Fe^{2+} concentrations of 300 and 350 mg/l, respectively. In addition at certain concentration in the range, Fe^{2+} increased the biomass yield as well (Wang et al., 2008). Furthermore Kim, et al. (2006) determined that the optimum sucrose concentration for the hydrogen production process was 30 g/l when they studied the effects of different sucrose concentrations on the production of hydrogen.

4.1.2 Objectives

- To investigate the formation of products and the optimum concentration of L-cysteine.HCl.H_2O within the growth medium for *Clostridium beijerinckii* to produce the highest cumulative biogas volume.

- To investigate the relationship between the metabolic shifts of organic acids produced and their individual concentrations on total volume of biogas produced.

4.2 Materials and Methods.

The fermentation trials for the one factor L.cysteine.HCl.H_2O medium design were all carried out aseptically, with all trials carried out in triplicate. *Clostridium beijerinckii* NCTC 13035 strain, storage condition and revival from storage were as described in sections 2.1.1 and 2.1.2. The starter cultures used to prepare the initial inocula were prepared as described previously (section 2.1.5) and the initial inoculums volume used was 10% of the volume of the culture medium used. The *C. beijerinckii* biomass was determined as described in section 2.3.2. The metabolic products were extracted and analysed as described in section

2.4.3.

The yield of biogas produced were determined by calculating the concentrations of products in the final biogas mixture assuming hydrogen and CO_2 were the only gasses produced from the head space of the culture medium in the syringe. Carbon recovery was calculated as the carbon contained in the organic metabolites without considering carbon contained in cell biomass. The concentration of each biogas produced from the total mixture of biogas was calculated using the ideal gas equation assuming the conditions were temperature of 37°C and pressure of 1 atmosphere. Theoretical hydrogen gas yield was calculated as the mole of hydrogen gas derived per mole of N-acetylglucosamine utilised. Furthermore it is the difference between the sum of hydrogen contained in the number or moles of organic acids per N-acetylglucosamine utilised and the number of moles of hydrogen contained per mole of N-acetylglucosamine (Appendix 4.2).

Analysis of variance was used as a statistical test for significant effects using the statistical software IBM Statistical Product and Service Solution package version 20 (SPSS). Bivariate correlation procedure computing with Pearson's correlation co-efficient was used to measure linear associations using the IBM SPSS package version 20.

4.2.1 Experimental Design

Based on the results derived from the Placket and Burman medium design trials with *C. beijerinckii,* a one-factor-at-a-time design was used to investigate the optimum concentration of L-cysteine.HCl.H$_2$O required to support maximal biogas production by *C. beijerinckii.* This was performed by varying the concentrations of L-cysteine.HCl.H$_2$O whilst keeping the concentration of other medium constituents constant. This design investigates one factor, while making the levels of other factors constant. When investigating a factor the level of the factor to be investigated is changed to analyze its effects on a response. In a one factor at a time design, information is gained about one medium constituent in each experimental run. This experimental procedure is repeated one after the other for all factors to be investigated. In the one factor at a time procedure, the response is optimized along the way. The steps involved when using the this design are explained below:

- A base line set of medium constituents levels (concentrations) are identified and executed, then the response is measured.

- The level each factor is changed and tested whilst keeping the other factors constant.

- This procedure is conducted for every factor.

Clostridium beijerinckii was cultured for 10 days in a 50 ml syringe. The basal medium comprised N-acetylglucosamine (15 g/l), bacteriological peptone (3 g/l), yeast extract (2 g/l), buffer capacity 0.2 M, FeSO$_4$.7H$_2$O (0.1 g/l), MgCl2 (0.43 g/l), at an initial pH 6.5 with L-cysteine.HCl.H2O at one of the following concentrations: 0 g/l; 0.5 g/l; 1 g/l; 1.5 g/l; 2 g/l; 2.5 g/l; 3 g/l; 3.5 g/l; 4g/l; or 4.5 g/l. This design was carried out with constituent variables producing significant effects and variable concentrations with highest effect responses in 10 different trial designs (C_1 to C_{10}) making a total of 30 batch fermentation runs.

4.3 Results

4.3.1 Profile of Biogas produced by *Clostridium beijerinckii* under different concentrations of L-Cysteine.HCl.H$_2$O.

The biogas production over time profiles for *Clostridium beijerinckii* in the different concentrations of L-cysteine.HCl.H$_2$O are shown in Figure 4.1.

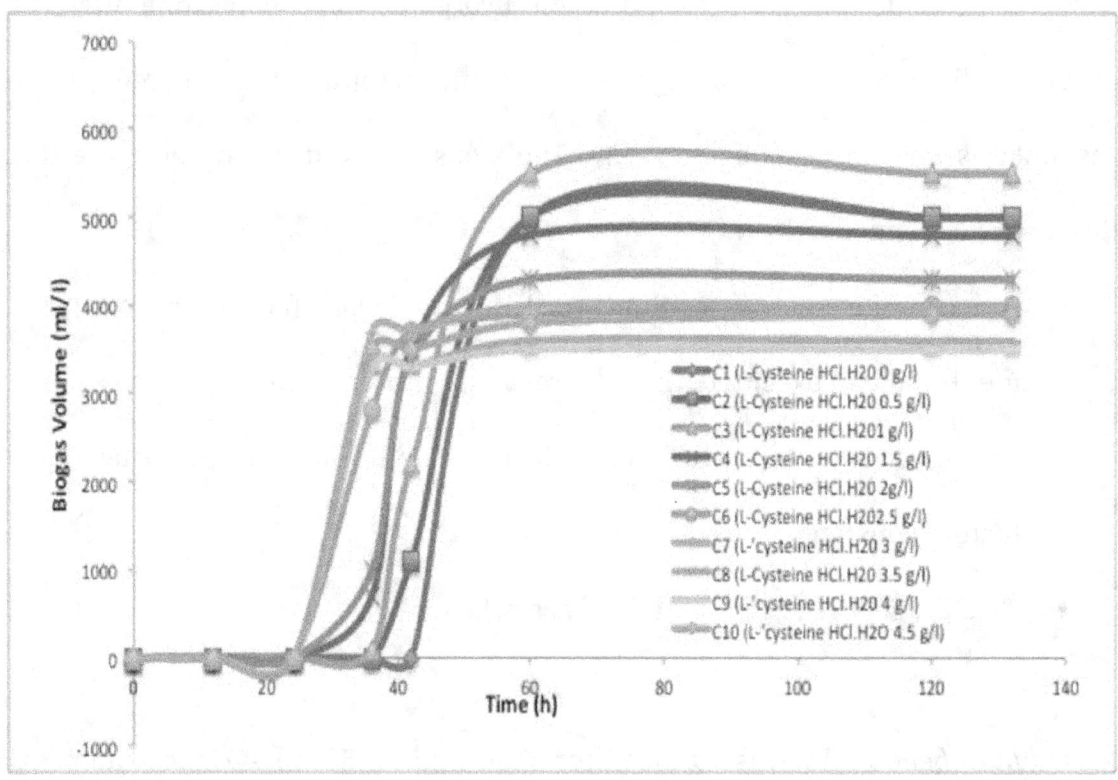

Figure 4.1 Profile of biogas production over time for *Clostridium beijerinckii* when supplied with different concentrations of L-cysteine.HCl.H$_2$O

    g/l = concentration in grams per litre.
    Biogas volume (ml/l) = Volume of biogas produced per litre of medium.
    C_1 to C_{10} = L-cysteine HCl.H$_2$O concentrations in grams per litre.
    h = time in hours.

132

n = 3

The medium formulations with the highest concentrations (3.5 g/l, 4 g/l, 4.5 g/l) of L-cysteine.HCl.H_2O were the first to initiate visible production of biogas, which occurred within 36 hours of incubation. On the other hand production of biogas was initiated at much later than 36 hours of incubation in the medium formulations with the lower concentrations of L-cysteine.HCl.H_2O.

The greatest volume of biogas (5466 ml/l) was produced when the L-cysteine.HCl.H_2O concentration was 1 g/l (Trial C_3; Figure 4.2). In contrast, the smallest volume of biogas (3533 ml/l) was produced when the concentration of L-cysteine.HCl.H2O was 4 g/l (Trial C_9; Figure 4.2). There was a significant difference (P = 0.001) between these sets of conditions (1g/l and 4 g/l of L-cysteine.HCl.H_2O). Surprisingly, the L-cysteine.HCl.H_2O negative control (0 g/l; Trial C_1; Figure 4.2) also produced a large volume of biogas (5000 ml/l), which was not significantly different (P > 0.303) from that produced when the L-cysteine.HCl.H_2O concentration was 1 g/l (Trial C_3), although it took a longer time after inoculation for biogas production to be initiated (Figure 4.1). Figure 4.2 shows the relationship between total volume of biogas and L-cysteine.HCl.H_2O concentration.

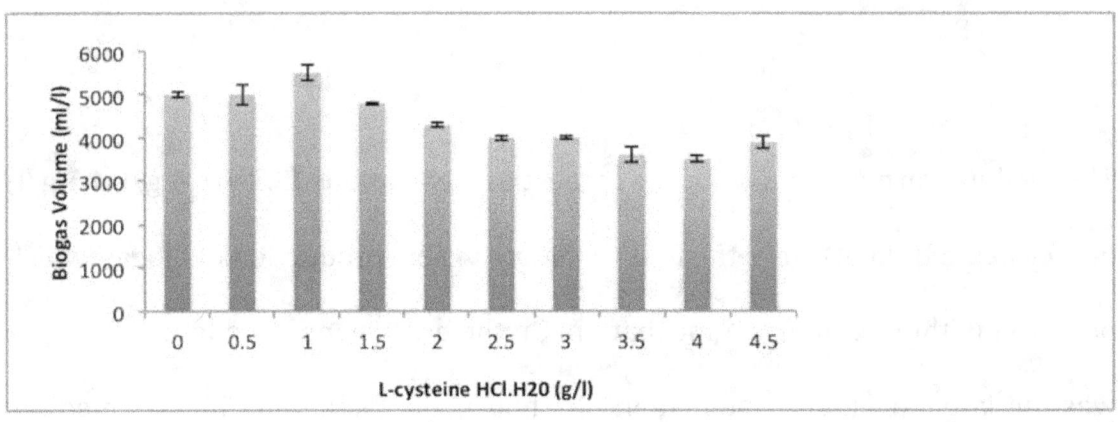

Figure 4.2 The relationship between the initial L-cysteine.HCl.H$_2$O concentration in the medium and total biogas volume produced from N-acetylglucosamine fermentation by *Clostridium beijerinckii*.

> g/l = concentration in grams per litre.
> Biogas volume (ml/l) = Volume of gas produced per litre of medium.
> n = 3

It is reasonable to conclude that as the L-cysteine.HCl.H$_2$O concentration was increased there was a significant (Within the range of 0 g/l to 4 g/l of L-cysteine.HCl.H$_2$O P < 0.001) trend towards a lower total volume of biogas produced. Between 4 g/l (C9) and 4.5 g/l (C10) of L-cysteine.HCl.H2O there was an insignificant difference (P > 0.717). Overall the test between the concentrations of L-cysteine.HCl.H$_2$O and the effect of total volume of biogas produced was considered significant (P = 0.000, Appendix 4, Table 4.4).

Figure 4.3 shows the relationship between the final cell density achieved by the cultures, and the initial L-cysteine.HCl.H$_2$O concentration present in the growth medium.

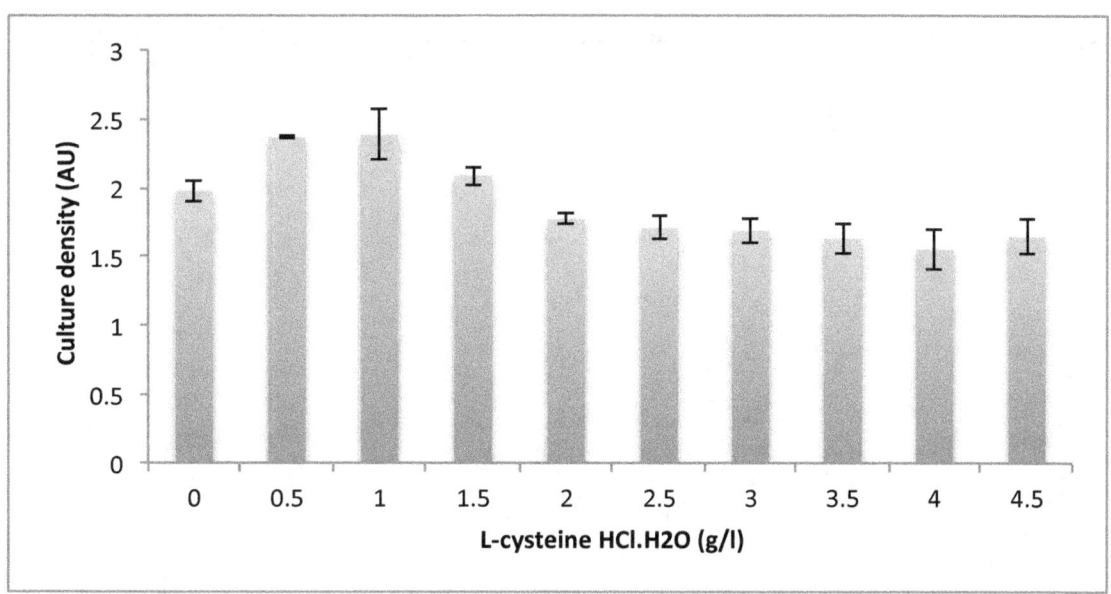

Figure 4.3 The relationship between the initial L-cysteine.HCl.H_2O concentration in the medium and the final cell density achieved by cultures of *Clostridium beijerinckii*

    g/l = concentration in grams per litre.
    AU = Absorbance Units at 560 nm.

Interestingly, there was a significant (P < 0.001) trend towards a reduction in the final cell density of the culture, as the initial concentration of L-cysteine.HCl.H_2O in the medium was increased, and this seemed to correlate (R^2 = 0.917, Appendix 4, Figure 4.1) with the decrease in biogas volume produced. The data demonstrate that an initial L-cysteine.HCl.H_2O concentration of 0.5 or 1 g/l resulted in cell densities of 2.369 and 2.383 AU, and total biogas volumes of 5000 and 5466 ml/l, respectively, whilst an initial L-cysteine.HCl.H_2O concentration of 4 g/l resulted in a final cell density of 1.554 AU and a total biogas volume of 3533 ml/l.

Figure 4.4 shows the relationship between the final pH values and the initial L-cysteine.HCl.H_2O concentrations of the culture medium. The initial pH of the

135

culture medium was 6.5, in all cases, and the buffering capacity of the medium comprised 0.2 M phosphate buffer.

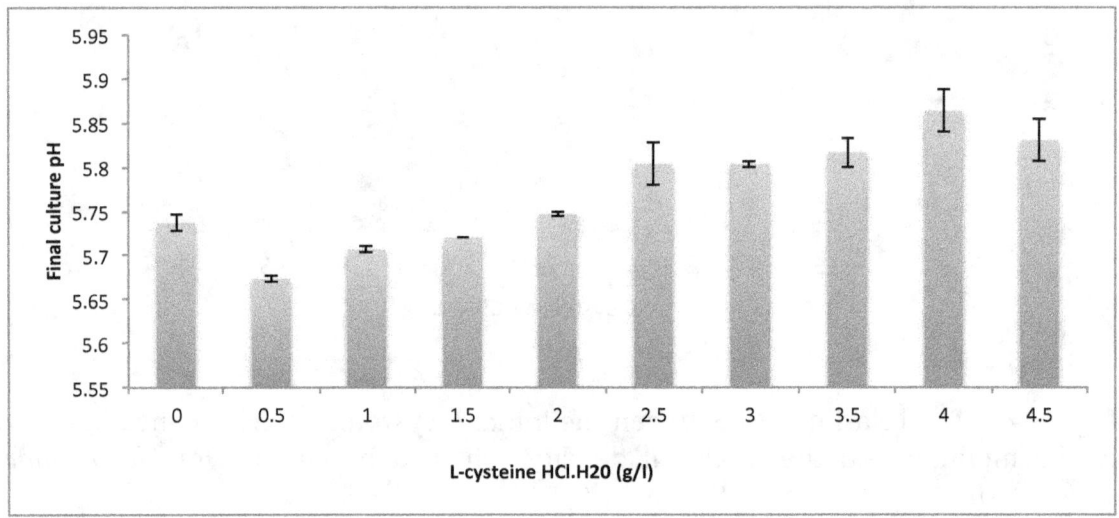

Figure 4.4 The relationship between the initial concentration of L-cysteine.HCl.H$_2$O and final the pH values of the growth medium in *Clostridium beijerinckii* cultures.

> g/l = initial concentration of L-cysteine.HCl.H$_2$O in the medium in grams per litre.

As the initial L-cysteine.HCl.H$_2$O concentration in the medium was increased, the final pH value of the medium also increased (figure 4.4) significantly (p = 0.000). However the change between the lowest initial L-cysteine.HCl.H$_2$O concentration (0.5 g/l) with the lowest final pH to high initial L-cysteine.HCl.H$_2$O concentration (4 g/l) and the highest final pH was small (pH 5.67 and 5.86) but considered significant (P < 0.001). For example, the lowest final pH of 5.67 was achieved in the trial using the lowest initial concentration of L-cysteine.HCl.H$_2$O at 0.5 g/l, whilst the highest final pH of 5.86 was achieved from an initial L-cysteine.HCl.H$_2$O concentration of 4.0 g/l.

4.3.2 The Effect of Initial L-cysteine.HCl.H₂O Concentration on Product Formation in Anaerobic Batch Cultures of *Clostridium beijerinckii*.

In order to investigate the effect of L-cysteine.HCl.H$_2$O concentrations on product formation from N-acetylglucosamine by *Clostridium beijerinckii*, batch culture trials were carried out in 50 ml syringes using culture medium formulations optimized form the other medium components during the Plackett and Burman experimental results described in Chapter 3. In these experiments, acetate and butyrate were the only organic acids detected. Butyrate was generally the most abundant organic acid produced in grams per liter from utilization of N-acetylglucosamine by the cultures in most of the L-cysteine.HCl.H$_2$O trials. Overall there were significant effects of L-cysteine.HCl.H$_2$O concentrations on butyrate ($P < 0.001$) and acetate ($P < 0.001$) production (Appendix 4, Table 4.4). Table 4.1 summarizes the GlcNAc consumed, and the final cell density, organic acids and concentration, and total biogas volume produced by *C. beijerinckii* cultures grown with the various initial L-cysteine.HCl.H$_2$O concentrations, after the formation of products and biogas had ceased. In the batch culture using 15 g GlcNAc/l medium, about 0.75 g/l, 1.62 g/l and 0.66 g/l of GlcNAc from trials C$_7$, C$_9$, C$_{10}$ respectively still remained in the culture liquid when fermentation ceased.

Table 4.1 The relationship between final cell density, N-acetylglucosamine consumed, biogas volume produced and the product yield in batch cultures of *Clostridium beijerinckii* at various concentrations of L-cysteine.HCl.H$_2$O.

L-Cysteine.HCl.H$_2$O (g/l)	Final Cell density AU/l	GlcNAc consumed (g/l)	Yield of product (g/l)		Biogas Volume (ml/l)
			Acetate	Butyrate	
C1 = 0	1.973	15.07	4.39	3.95	5000
C2 = 0.5	2.369	14.88	4.28	4.18	5000
C3 = 1	2.383	14.69	4.41	4.31	5467
C4 = 1.5	2.088	14.78	4.50	4.32	4833
C5 = 2	1.775	15.06	4.23	4.56	4300
C6 = 2.5	1.709	14.94	3.96	4.81	3967
C7 = 3	1.687	14.51	3.70	5.00	3967
C8 = 3.5	1.629	14.84	3.75	5.03	3567
C9 = 4	1.555	13.24	3.42	4.50	3533
C10 = 4.5	1.645	14.39	3.81	4.80	3867

g/l = concentration in grams per litre.
Biogas volume (ml/l) = Volume of gas produced per litre of medium.
C_1 to C_{10} = L-cysteine HCl.H$_2$O concentrations in grams per litre.
AU = Absorbance Units (measured at 560 nm)

Comparing the growth of *C. beijerinckkii* and the yield of its products with biogas produced, it was observed in table 4.1 that as acetate concentration decreased with increasing concentration of L-cysteine.HCl.H$_2$O, the concentration of butyrate increased gradually. In other words increase in initial concentrations of L-cysteine.HCl.H$_2$O correlated (negatively) significantly with a decrease in acetate concentrations (R^2 = 0.838, P < 0.001, Appendix 4, Table 4.10) and increases in butyrate concentrations (R^2 = 0.785, P < 0.001, Appendix 4, Table 4.11). Decreases in total biogas volume correlated significantly with decreases in acetate concentrations (R^2 = 0.830, P < 0.001, Appendix 4, Table 4.12) whilst decrease in total biogas volume correlated negatively with increase in butyrate concentrations (R^2 = 0.730, P < 0.001, Appendix 4, Table 4.13).

The yield coefficients (moles of product per moles of substrate consumed), the carbon recovery, the ratio acetate produced to butyrate produced, and the theoretical yield of hydrogen gas from anaerobic fermentation N-acetylglucosamine by *C. beijerinckii* in is shown in table 4.2. The final yield coefficient of each product changed according to initial L-cysteine.HCl.H$_2$O concentrations in the medium design. The carbon recovery was calculated in percentages as the total amount in moles of carbon contained in metabolized acetate and butyrate produced per mole of GlcNAc consumed as shown in table 4.2.

Table 4.2 The relationship between product yield coefficient (moles of product per mole of substrate consumed), percent carbon recovery from the GlcNAc source, the ratio of acetate to butyrate, and the theoretical yield of hydrogen gas from batch cultures of *Clostridium beijerinckii* at the various initial concentrations of L-cysteine.HCl.H$_2$O tested

Cysteine trials (g/l)	GlcNAc consumed (g/l)	Yield coefficient of product (mol of product/mol GlcNAc consumed)		Carbon recovery from GlcNAc (%)	Acetate/ butyrate (mol/mol)	Theoretical yeild of Hydrogen gas. (mol of H2/mol GlcNAc)
		Acetate	Butyrate			
C1 = 0	15.07	1.09	0.67	61	1.64	3.53
C2 = 0.5	14.88	1.08	0.71	63	1.51	3.39
C3 = 1	14.69	1.13	0.75	65	1.51	3.20
C4 = 1.5	14.78	1.14	0.74	66	1.54	3.19
C5 = 2	15.06	1.05	0.77	65	1.37	3.23
C6 = 2.5	14.94	0.99	0.82	66	1.21	3.14
C7 = 3	14.51	0.95	0.88	68	1.09	3.00
C8 = 3.5	14.84	0.95	0.86	67	1.10	3.07
C9 = 4	13.24	0.97	0.86	67	1.12	3.02
C10 = 4.5	14.39	0.99	0.85	67	1.17	3.04

g/l = concentration in grams per litre.
C$_1$ to C$_{10}$ = the initial concentration of L-cysteine HCl.H$_2$O in the growth medium in grams per litre.
% = Percentage.

The carbon recovery from the fermentations described here increased with increasing initial L-cysteine.HCl.H$_2$O concentration in the medium. However, the opposite was the case for the theoretical hydrogen gas yield, which was reduced with increasing initial L-cysteine.HCl.H$_2$O concentration in the medium. Theoretically, the greatest yield of hydrogen gas (at 3.53 mol of H$_2$/ mol GlcNAc) are from cultures grown in an initial concentration of 0 g/l L-cysteine.HCl.H$_2$O, whilst the trial with the smallest theoretical yield of hydrogen gas (at 3 mol of H$_2$/ mol GlcNAc) were from cultures grown in an initial concentration of 3 g/l L-cysteine.HCl.H$_2$O. The yield coefficient of acetate in terms of moles of acetate produced per mol of GlcNAc consumed was reduced significantly (P = 0.00, Appendix 4, Table 4.6) from 1.09 to 0.99 moles per moles of N-acetylglucosamine with increasing initial concentrations of L-cysteine.HCl.H$_2$O, but the theoretical yield of hydrogen was reduced from 3.53 to 3.04 moles of H$_2$ per mole of N-acetylglucosamine. In contrast, the yield coefficient for butyrate, in terms of moles of butyrate produced per mole of GlcNAc consumed increased significantly (P < 0.001, Appendix 4, Table 4.6) with an increasing initial concentration of L-cysteine.HCl.H$_2$O in the medium, and at the same time the theoretical yield of hydrogen produced reduced. In other words the molar ratio of acetate to butyrate was reduced as the initial concentration of L-cysteine.HCl.H$_2$O was increased, and at the same time the theoretical yield of hydrogen was reduced whilst the carbon recovery was increased.

Table 4.3 shows the relationship between the initial L-cysteine.HCl.H$_2$O concentrations used and the yield of products from the consumption of N-acetlglucosamine by *Clostridium beijerinckii* cultivated in an anaerobic batch culture system. Whilst Table 4.4 shows the relationship between the initial L-cysteine.HCl.H$_2$O concentration, and the yield coefficient for moles of biogas produced per mole of N-acetylglucosamine utilised in an anaerobic batch culture system.

Table 4.3 The relationship between N-acetylglucosamine consumption, the product yield, and the total biogas volume produced from batch cultures of *Clostridium beijerinckii* at various concentrations of L-cysteine.HCl.H_2O.

| Cysteine trials (g/l) | GlcNAc Consumed (g/l) | Yield of products | | | | Biogas Volume (ml/l) |
		Acetate (g/l)	Butyrate (g/l)	H_2 gas (ml/l)	CO_2 gas (ml/l)	
C1 = 0	15.07	4.39	3.95	4320	680	5000
C2 = 0.5	14.88	4.28	4.18	4355	645	5000
C3 = 1	14.69	4.41	4.31	4870	597	5467
C4 = 1.5	14.78	4.50	4.32	4240	594	4833
C5 = 2	15.06	4.23	4.56	3691	609	4300
C6 = 2.5	14.94	3.96	4.81	3376	591	3967
C7 = 3	14.51	3.70	5.00	3408	559	3967
C8 = 3.5	14.84	3.75	5.03	2992	575	3567
C9 = 4	13.24	3.42	4.50	2971	563	3533
C10 = 4.5	14.39	3.81	4.80	3301	566	3867

g/l = concentration in grams per litre.
C_1 to C_{10} = the initial L-cysteine HCl.H_2O concentration in the medium in grams per litre.
ml/l = millilitres per litre
% = Percentage

Table 4.4 The yield coefficient for moles of biogas produced per mole of N-acetylglucosamine consumed in anaerobic batch cultures of *Clostridium beijerinckii*, using the ideal gas law assuming conditions of temperature at 37°C and pressure at 1 atm.

| Cysteine trials (g/l) | Yield coefficient of product (biogas) per GlcNAc consumed (mol/mol) | | | H_2 gas expelled (%) | CO_2 gas expelled (%) |
	Biogas	H_2	CO_2		
C1 = 0	2.90	2.50	0.39	86	14
C2 = 0.5	2.90	2.52	0.37	87	13
C3 = 1	3.17	2.82	0.35	89	11
C4 = 1.5	2.80	2.46	0.34	88	12
C5 = 2	2.49	2.14	0.35	86	14
C6 = 2.5	2.30	1.96	0.34	85	15
C7 = 3	2.30	1.97	0.32	86	14
C8 = 3.5	2.07	1.73	0.33	84	16
C9 = 4	2.05	1.72	0.33	84	16
C10 = 4.5	2.24	1.91	0.33	85	15

g/l = concentration in grams per litre.

C_1 to C_{10} = the initial L-cysteine.HCl.H_2O concentration in the medium in grams per litre.

% = Percentage

The yield of hydrogen gas (at 4941 ml/l) and of CO_2 (at 525 ml/l), equivalent to 89% H_2 and 11% CO_2, was greater at an initial L-cysteine.HCl.H_2O concentration of 1 g/l than at any other initial L-cysteine.HCl.H_2O concentration ($P = 0.00$, Appendix 4, Table 4.5) but was significantly greater than trials with L-cysteine.HCl.H_2O concentrations ranging from 2 g/l to 4.5 g/l. The butyrate yield showed a correlation with the initial L-cysteine.HCl.H_2O concentration, but an inverse correlation with acetate, hydrogen gas, CO_2 gas and total biogas yield (Table 4.3).

4.4 Discussion

4.4.1 Profile of Biogas Production with the Different Initial concentrations of L-cysteine.HCl.H_2O.

The *Clostridium beijerinckii* cultures grown in the presence of higher initial L-cysteine.HCl.H_2O concentrations initiated visible production of biogas sooner than those grown in the presence of lower concentration of L-cysteine.HCl.H_2O. It is probable that this resulted from L-cysteine.HCl.H_2O helping to make the culture medium a more strongly reducing environment, due to scavenging of any trace amounts of oxygen present in the medium (Sinji, et al., 1998) thereby helping to initiate cell growth more rapidly. Inhibition of clostridial growth by the presence of oxygen in the growth medium has been observed previously and was suggested to be because clostridia lack the mechanisms for removal of

142

oxygen derivatives and reactive by-products such as hydrogen peroxide, hydroxyl radicals and superoxide (Sinji, et al., 1998). These reactive by-products are harmful to bacterial cell components such as phospholipids, DNA and other biomolecules (Elisa et al., 2000). Aerobic organisms have developed mechanisms to protection against oxidative stress, with enzymes such as catalase and superoxide dismutase (Elisa et al., 2000). However, the lack of a catalase enzyme is known to cause aerointolerance in Clostridium.

Nevertheless, it is known that *Clostridium butyricum* is capable of consuming oxygen (Sinji, et al., 1998). In cultures of Clostridium exposed to medium contaminated with oxygen, it was shown that growth was interrupted whilst oxygen was consumed, but re-commenced once a low redox potential was achieved (Sinji, et al., 1998). It is plausible that the period of time required for *C. butyricum* to consume the oxygen in the medium before growth is initiated may be part of the reason for the lag that it demonstrates before fermentation commences. On the other hand, it is not currently known whether *C. beijerinckii* has the same oxygen consumption mechanisms as *C. butyricum*. However when higher concentrations of L-cysteine.HCl.H$_2$O were present in the medium lag time before fermentation was initiated was shorter. Sim and Kamaruddin (2008) have also observed that L-cysteine was required for the growth of *Clostridium beijerinckii* and suggested that this might result in the redox potential of the medium being reduced by L-cysteine to permit cell growth and fermentation to be initiated. The current experiments (shown in Figure 4.1) demonstrated that cultures of *C. beijerinckii* produced biogas earlier when incubated with higher initial concentrations of L-cysteine.HCl.H$_2$O than those with a lower concentration of L-cysteine.HCl.H$_2$O, which supports the observations of Sim and Kamaruddin (2008). In the current experiments, as L-cysteine.HCl.H$_2$O concentrations increases there were shorter lag times and fermentation of biogas duration. However the high initial L-cysteine.HCl.H$_2$O concentrations in the growth medium, which help to initiate biogas production sooner, than at low concentrations, does not necessarily equate to a large total volume of biogas produced. There was a substantial difference between the early and late initiations of biogas relating to high and low concentrations of L-cysteine.HCl.H$_2$O respectively. This also proves the relationship between total biogas volume and different concentrations of cysteine.

Yaun et al. (2008) studied the enhancement of hydrogen production by anaerobic mixed cultures in the presence of L-cysteine during dark fermentation. Similar results to those reported in this work were obtained by Yaun et al. (2008) with L-cysteine concentrations ranging from 0.012 to 0.12 g/l (0.1 to 1.0 mM), the lag times decreased noticeably to 1.2–4.0 h, which was shorter than 8.4 h of the negative control whilst maximum H$_2$ production rate of experiments increased from 9.0 to 10.2 ml/h. In this report the lag time after inoculation for most of the L-cysteine positive *C. beijerinckii* cultures was about 36 h, whilst for the L-cysteine negative control culture it was 60 h. In addition, in the report by Yaun et al. (2008) the duration of the fermentation yielding H$_2$ was reduced from 69h for the negative control culture to 50 h at initial L-cysteine concentration ranging from 0.012 to 0.12 g/l of L-cysteine. In this report, biogas production was also initiated earlier in the fermentation as the initial L-cysteine.HCl.H_2O concentration was increased, which strongly suggests that L-cysteine.HCl.H_2O made the medium more supportive of biogas production activity by *C. beijerinckii* during the fermentations.

The maximum volume (5467 ml/l) of biogas produced in the experiments reported in this work was achieved at an initial L-cysteine.HCl.$H_2$0 concentration of 1 g/l (equivalent to a cysteine concentration of 0.69 g/l) and a hydrogen yield of 2.82 moles of H_2 per mole of GlcNAc (Considering ideal gas conditions Table 4.4). Similarly, Yuan, et al. (2008) observed that a maximum total hydrogen volume of 199 ml (2487.5 ml/l) at 35°C and a maximal hydrogen yield of 3.10 mole H_2 per mole sucrose were obtained in fermentation experiments with the addition of an initial 0.6 mM L-cysteine (0.073 g/l). The increase in biogas in both the work of Yuan et al. (2008) and the current experiments is likely to have been a result of lowered redox potential in the fermentation medium brought about by the addition of L-cysteine, which led to the production of biogas by *C. beijerinckii* in the fermentation process sooner than in cultures not supplemented with L-cysteine.

It was also observed in the current experiments that as the L-cysteine concentration increased there was a reduction in final biogas volume. Yuan et al. 2008) also demonstrated that when the initial concentration of L-cysteine in the medium was increased above 0.2 mM (0.024 g/l) at 30°C the volume of hydrogen gas produced was reduced. Since the greatest of volume biogas (5000 and 5466 ml/l) was produced from medium supplemented with either no or very little L-cysteine.HCl.H$_2$O (0 and 1 g/l, respectively), whilst the smallest volume of biogas (3533 ml/l) was produced from medium supplemented with 4 g/l L-cysteine.HCl.H$_2$O, it is reasonable to conclude that as cysteine concentration is increased there was a tendency towards smaller final biogas volumes produced. Decreased bioactivity of, and biogas yield from, *C. beijerinckii* may result from the ability of L-cysteine to absorb metal ions with its two carboxyl groups, two amido groups and two sulfur atoms per molecule, meaning that surplus L-cysteine in the medium might combine with these components of the medium and so make them unavailable for metabolism (Yu et al., 2007).

There were relationships between final cell density and initial cysteine concentration in the current experiments, as the initial cysteine concentrations in the medium were increased, the final cell density achieved by *Clostridium beijerinckii* was decreased significantly (P < 0.001). It cannot be concluded from the current work that an increase in cysteine concentration was toxic to *C. beijerinckii*, however Kari et al. (1971) have reported previously that high levels of cysteine can inhibit growth of *E. coli* by two different processes. Firstly, cysteine may obstruct biosynthesis of the essential amino acids threonine, isoleucine, leucine and valine and secondly it may have an effect on the function of membrane-bound enzymes. Furthermore, Carlsson et al. (1978) have reported that hydrogen peroxide is formed in the presence of cysteine in a metal ion-catalysed reaction in the presence of oxygen and that the hydrogen perioxide produced was toxic to *Peptostreptococcus anaerobius* VPI 4330-1. Moreover, since cysteine can also reduce the availability to bacteria of certain nutrients during fermentation, this could also have resulted in a lower final cell density at higher initial L-cysteine concentrations. It would be practical to relate high optical cell density with final biogas volume as trials C_2 (0.5 g/l of L-cysteine.HCl.H$_2$O) and C_3 (1 g/l of L-cysteine.HCl.H$_2$O) with optical cell densities of 2.369 AU and 2.383 AU had high final biogas production volume of 5000 ml/l and 5466 ml/l respectively while C_9 (4 g/l of L-cysteine.HCl.H$_2$O) trial with the lowest cell density of 1.554 AU had the lowest final biogas volume of 3533 ml/l.

The production of organic acids was responsible for a decrease in pH value in each of the L-cysteine.HCl.H$_2$O medium formulation trial experiments. There was a negative significant correlation (R^2 = 0.873, P < 0.001, Appendix 4, Table 4.8) between the culture final pH and acetate and a positive significant correlation (R^2 = 0.623, P < 0.001, Appendix 4, Table 4.9) between culture final pH and butyrate concentration produced. This indicates a stronger correlation between acetate and culture final pH than butyrate and culture final pH. The final pH value decreased as the acetate concentration increased. One possible explanation for slightly decreased final pH values might be that acetate conceals the pH effect of butyrate. Butyric acid is a weak acid with a pKa (acid dissociation constant) of 4.82, while acetic acid has pKa of 4.76. Thus it can be concluded that lower final acetate yields (Table 4.1) would lead to higher final pH values (Figure 4.4). Furthermore as 2 moles of acetate are produced per mole of glucose, 1 mole of butyrate is produced per mole of glucose, thus there are twice as many potential sources of protons from dissociation (Equation 4.1 and 4.2). These may be the reasons why the effect of acetate on pH value is more evident in the culture medium than butyrate.

In addition to changes in the acetate and butyrate concentrations, another possible explanation for increased pH with increased initial cysteine concentrations could be reductions in the dissolved CO_2 concentrations in the liquid culture medium. Since dissolution of carbon dioxide in aqueous culture media increases the acidity of and lowers the pH, then decreased production of CO_2 by the cultures in the presence of higher L-cysteine.HCl.H$_2$O might have permitted the pH to rise. It can be assumed that a proportion of CO_2 produced would be dissolved in the culture medium, making it more acidic. Therefore as dissolved CO_2 increases pH value decreases and as dissolved CO_2 decreases final pH value increases.

4.4.2 Effect of Cysteine (L-cysteine.HCl.H$_2$O) Concentration on Product Formation in Anaerobic Batch Cultures of *Clostridium beijerinckii*.

In the experiments described in this work, acetate and butyrate were the only organic acids detected. In terms of grams per litre, butyrate was generally the most abundant organic acid produced from N-acetylglucosamine in most of L-cysteine.HCl.H$_2$O medium formulation trials. According to equations 4.1 and 4.2 production of acetate and butyrate results in production of hydrogen gas.

Acetic acid production

$$C_6H_{12}O_6 + 2H_2O \rightarrow 2CH_3COOH + 2CO_2 + 4H_2 \quad\text{.....................(4.1)}$$

Butyric acid production

$$C_6H_{12}O_6 \rightarrow CH_3CH_2CH_2COOH + 2CO_2 + 2H_2 \quad\text{.......................(4.2)}$$

The reactions shown in equations 4.1 and 4.2 demonstrate a yield of 4 moles of hydrogen from each mole of glucose when acetate is produced, but only 2 moles of hydrogen is produced from each mole of glucose when butyrate is produced (Nandi & Sengupta, 1998). Yaun, et al. (2008) stated that the production of butyrate and acetate are in accordance with the trend of hydrogen production, when studying metabolite distribution from the effect of L-cysteine on dark fermentative hydrogen production using mixed cultures. Moreover it was found in that study that acidogenic bacteria, producing butyrate or acetate, were also the major producers of hydrogen gas (Yuan et al., 2008).

There is an inverse relationship between the proportion of acetate or butyrate produced, and the total amount of hydrogen gas evolved at different initial L-cysteine.HCl.H$_2$O concentrations (Table 4.2). When more acetate is produced compared to butyrate there is a higher theoretical yield of hydrogen gas. The experiments described in this work demonstrate that as the initial cysteine concentration was increased, the acetate concentration and biogas yield were decreased, whilst the butyrate concentration was increased. This indicates that medium formulations containing higher initial concentrations of L-cysteine do not favour production of biogas by *C. beijerinckii*. From equation 4.1 it is predicted that a greater volume of hydrogen gas (4 mol) will be produced when acetate is a product of fermentation, than will be produced (2 mol) when butyrate is a product of fermentation. From the equations (4.1 and 4.2) it also concludes the results attained in this report, as biogas volume reduced it correlated with a decrease in acetate concentration.

L-cysteine.HCl.H_2O made the medium more conducive for the initiation of biogas production as biogas production was also initiated earlier in the fermentation. This was as a result of lowered redox potential in the fermentation medium. However the ability of L-cysteine to absorb metal results in combining effect with components of the medium and so makes them unavailable for metabolism (Yu et al., 2007) as well as carbon utilization and organic acid production. This is concluded as a reason for lower biogas volume with high cysteine concentration.

Acetate is usually thought as the final product produced during anaerobic fermentation (Sylvia et al., 2004). It will seem that when acetate is produced there is a demand for more energy as N-acetylglucosamine is metabolized to the lowest carbon-containing product. In other words carbon energy contained in N-acetylglucosamine is used up to the lowest carbon-containing product. But when more butyrate is produced more carbon energy is contained in the by-product, suggesting energy utilization and demand by the clostridium bacteria is reduced. This might be the reason why there were unconsumed GlcNAc in trials with high concentrations of cysteine (C7, C9, C10).

Furthermore there is a greater carbon energy contained in butyrate, than acetate and as cysteine concentrations increase there were more butyrate left in the culture medium than acetate. If there is residual carbon source left in the medium, then there has been a reduction in N-acetylglucosamine utilised and a concomitant reduction in the carbon energy released by the cells and also cell density. It can be inferred that there was a decrease in DNA and RNA synthesis due to a reduction in final overall cell density (table 4.1). A reduction in N-acetylglucosamine metabolism would also have the effect of reducing the availability of ATP and thus the reducing power available to the cell, since GlcNAc was the main carbon source available, which will eventually result in energy starvation for the *C. beijerinckii*. This observation also explains the lower final cell density leading to reduced total yield of biogas with increased initial cysteine concentrations. Yuan et al. (2008) reported that total hydrogen volumes and hydrogen yields produced demonstrated a rapid increase and then a slow decrease correlating with the increase of L-cysteine concentration and then concluded that cysteine shifts the metabolic pathways of anaerobic mixed cultures to produce high concentrations of acetate and butyrate. Sim and Kamaruddin (2008) suggest that cysteine initiates cell growth, this indicates that cysteine may not be toxic to the *Clostridium* cells instead it may have changed the metabolic pathway and energy utilization.

Table 4.2 further explains this as carbon recovery from N-acetylglucosamine increases with the increase in cysteine concentration indicating that more carbon containing metabolite was detected with the increase in cysteine concentration. Following on from the increase in carbon recovery at higher initial L-cysteine concentrations, it was calculated that a smaller volume of CO_2 gas was evolved (table 4.3) because of this. The final yield coefficient of each product also changed according to the initial cysteine concentration used. Equations 4.1 and 4.2 predict that a high acetate to butyrate molar ratio gives a higher theoretical yield of hydrogen gas than a low acetate to butyrate ratio does, and this was matched by the experimental data which demonstrated greater production of hydrogen gas by those cultures which also produced a higher ratio of acetate to butyrate.

The pH of the fermentation culture medium may also play an important role in controlling which organic acids are produced. Zhu and Yang (2004) studied effect of pHs between 5.0 and 6.3 on xylose fermentation pathway shift by *Clostridium tyrobutyricum* and observed that at a pH of 6.3 butyrate was the main product whilst at pH below 5.7 acetate and lactate were the main products, with little butyric acid produced. The change from butyrate production pathways to lactate and acetate production pathways during fermentation was linked to changes in key enzyme activities. Butyrate formation is controlled by the enzyme phosphotransbutyrylase (PTB) whilst the conversion of lactate to pyruvate is catalyzed by the NAD-independent lactate dehydrogenase (iLDH). At a pH of 6.3 the activities of these two enzymes were higher in bacterial cells that produce mainly butyrate. In contrast, Phosphotransacetylase (PTA), which is the main enzyme controlling the production of acetate, and lactate dehydrogenase (LDH), which is the enzyme that catalyzes the conversion of pyruvate to lactate demonstrated higher catalytic activities when bacterial cells were cultured at a pH of 5.0. The different rates of xylose metabolism shown by the cells at the different pH values, suggested that the balance in NADH was the main factor controlling the metabolic pathways used for fermentation (Zhu and Yang, 2004).

However, it would seem that there is an important relationship between acetate and butyrate production, Jaros, Rova and Berglund (2011) studied the effect of acetate upon the fermentative production of butyrate by *Clostridium tyrobutyricum.* There was a lag of 45 and 118 hours, respectively, at initial acetate concentrations of 17.6 g/l and 26.3 g/l in the growth medium, based upon cell biomass and butyrate yield, and sugar consumption. When high concentrations of acetate were added to the initial medium re-uptake of acetate was exhibited and led to greater butyrate production. After the lag phase, there was inhibition in acetate production, which increased the final butyrate yields significantly by 12.6%. Furthermore it has been demonstrated that butyrate production can be stimulated by the addition of acetate to the growth medium for *C. thermobutyricum* (Canganella et al., 2002). In those experiments 14C-labelled acetate was supplied to the clostridial culture and its metabolism resulted in the formation of 44% labeled butyrate.

During investigations of the enzymes associated with formation of acetate and butyrate by *Clostridium acetobutylicum* (Hartmanis and Gatenbeck, 1984). It was found that there was a decrease in the specific enzyme activity phosphate acetyltransferase and acetate kinase, which help in the formation of acetate from acetyl coenzyme A, at the beginning of the solvent production stage. On the other hand, thiolase, β-hydroxybutyryl coenzyme A dehydrogenase and crotonase enzymes were expressed, and maximal enzyme activities were achieved, after growth had ended. Only low concentration levels of butyryl coenzyme A dehydrogenase enzyme activities were detected. The phosphate butyryltransferase enzyme activities decreased below the detection limit after 20 h of incubation. At this time butyrate production ceased and it was suggested that there was a metabolite production pathway change at the initiation of butanol production (Hartmanis and Gatenbeck, 1984).

Similarly, it was demonstrated that butyrate could be taken up by cells and converted to butanol during the *Clostridium acetobutylicum* solvent production phase (Hartmanis et al., 1984). In those experiments it was found that the activities of acetate phosphotransacetylase, phosphate butyryltransferase and acetate kinase declined to low levels at the initiation of solvent production. In addition, no short-chain acyl-CoA synthetase enzyme activity or butyryl phosphate reducing activity was detected. It was suggested by those authors that in *C. acetobutylicum* acetoacetyl-CoA, acetate or butyrate CoA-transferase was responsible for uptake and activation of acetate and butyrate. The main enzyme associated with the uptake of acetate and butyrate is acetoacetate decarboxylase, which is induced late in the fermentation stage and influences the transferase reaction leading to formation of acetoacetate. This means that when solvent production is initiated the production of acetate and butyrate ceases. However in this experiment there were no solvents detected or solvents that may have been produced were below detection limits. Although *C.beijerinckii* has the ability to

154

produce ethanol as shown in Table 3.5 and 3.10 in small quantities.

4.5 Conclusions

Cultures of *Clostridium beijerinckii* grown in medium formulations with a high initial concentration of L-cysteine.HCl.H_2O were the first to initiate visible production of biogas. When the growth of *C. beijerinckkii* was compared with the product and biogas yield, it was noted that a decrease in acetate concentration was significantly correlated with an increase in the initial L-cysteine.HCl.H_2O concentration, and these were both significantly correlated with an increased butyrate concentration.

The application of this optimization strategy significantly increased the final biogas volume produced from the utilization of N-acetlglucosamine by *Clostridium beijerinckii* from 3183 ml/l to 5467 ml/l using the one factor at a time L-cysteine.HCl.H_2O concentration optimization method. This means that despite purging the medium with nitrogen to remove oxygen residual dissolved oxygen or oxidizing agents in non-gaseous form were scavenged by cysteine to improve the biogas production. However an ideal concentration of cysteine is needed to make the medium reducing to initiate growth of *C. beijerinckii* as high concentrations of cysteine will reduce overall biogas produced; consequently an L-cysteine.HCl.H_2O concentration of 1 g/l would be recommended to get optimal growth and maximum total volume of biogas produced.

Chapter 5
Method Of Steepest Ascent

5.1.1 Background

To optimize the expression of a desired bacterial response requires that the most suitable conditions for the controllable factors, such as nutrient availability and environmental conditions, be determined. In the current experiments, the key objective is to determine the conditions close to the vicinity of optimum that will result in optimal biogas production. It can often happen that although initial analyses will indicate broadly the best conditions for a desired product, they still do not indicate the true optimum. Consequently, in such situations the objective of further analyses is to move closer to the true optimum.

With this in mind, it was decided to use the Method of Steepest Ascent to optimize the medium design for biogas production further. The Method of Steepest Ascent is a technique that works on the principle of working through a logical series of changes to the medium composition so as to move along the "path of steepest ascent", thereby getting closer to the vicinity of optimum. In other words the path of steepest ascent describes those changes to medium components that give the greatest increase in the desired response by the bacterial culture. However, if a minimal response is the desired result, then the technique is known as the Method of Steepest Descent. The factors initially identified by the Plackett and Burman medium design trials can be further investigated and fine-tuned using this method.

The Method of Steepest Ascent begins from the design of the factorial design improving the first-order polynomial model and stops when there is no further improvement in the response. This indicates that the area of optimum response of that condition has been reached (Kuehl, 2000).

After the medium design experiments have been carried out, the results should normally initially display improving response values. However, at some point the response will start to decline, after which the medium composition is no longer favourable for the desired response. The method of steepest ascent has been used to optimise three key factors – xylose, $FeSO_4$ and peptone – for the production of hydrogen by *Enterobacter species* CN1 (Long et al., 2010). This technique was applied to reach the area of optimum of the three significant factors.

5.1.2 Objectives

- To determine the medium component conditions that yield optimal biogas produced by *C. beijerinckii* and *C. paraputrificum* using the variables identified by the Plackett and Burman design experiments described in chapter 3. This is done by varying the medium constituents along the vicinity of optimum influence to produce the maximum cumulative biogas.

5.2 Materials and Method

Clostridium beijerinckii NCTC 13035 and *Clostridium paraputrificum* NCTC Y1833 strains, storage conditions and revival for this design are as described in section 2.1.1 and 2.1.2. Starter cultures were used as the intial inocula for the experiments as described in section   2.1.5. Clostridial biomass was measured

158

as described in section 2.3.2. The medium designs were evaluated independently in triplicate. Metabolic products were analyzed as described in section 2.4.3.

Carbon recovery was calculated as the carbon contained in the organic metabolites without considering carbon contained in cell biomass. Analysis of variance was used as a statistical test for significant effects using the statistical software IBM Statistical Product and Service Solution package version 20 (SPSS). Bivariate correlation procedure computing with Pearson's correlation co-efficient was used to measure linear associations using the IBM SPSS package version 20.

5.2.1 Experimental Design

The Method of Steepest Ascent was used to optimize fully the medium component variables. This was based upon regression analysis of the results derived from the Plackett and Burman medium design trials described previously.

Clostridium beijerinckii and *C. paraputrificum* were cultured for 8 days in a 50 ml syringe to collect biogas and samples for metabolic product analysis were taken after fermentation had ceased.

The medium design trials for *C. beijerinckii* described in this chapter were carried out by increasing the initial concentration of N-acetylglucosamine, whilst simultaneously reducing the concentration of $FeSO_4.7H_2O$ and also the initial pH of the medium through the trial   design as shown in table 5.1. For the *C.*

159

paraputrificum medium design trials, the concentrations of N-acetylglucosamine and magnesium chloride were increased, whilst the concentration of L-cysteine.HCl.H_2O was simultaneously reduced as shown in table 5.2 for each culture trial. The intended outcome of these medium design trials was to optimise the volume of biogas produced by the bacteria from the fermentation of N-acteylglucosamine.

The experiments in triplicate using batch cultures with a total of 5 different medium compositions were performed for *C. beijerinckii* (Table 5.1) and in triplicate using batch cultures with a total of 6 different medium compositions were performed for for *C. paraputrificum* (Table 5.2).

Table 5.1 Medium design compositions for Method of Steepest Ascent trials with *C. beijerinckii*

Trial Number	N-acetylglucosamine (g/l)	FeSO$_4$.7H$_2$O (g/l)	Initial pH
1	12	0.10	6.5
2	17	0.08	6.3
3	22	0.06	6.0
4	27	0.04	5.7
5	32	0.02	5.5

g/l = grams per litre.

Table 5.2 Medium design compositions for Method of Steepest Ascent with *C. paraputrificum.*

Trial Number	N-acetylglucosamine (g/l)	L-cysteine.HCl.H$_2$O (g/l)	MgCl$_2$ (g/l)
1	12	0.75	0.30
2	17	0.50	0.35
3	22	0.35	0.40
4	27	0.25	0.45
5	32	0.15	0.50
6	37	0.10	0.55

g/l = grams per litre.

All of the other medium constituents, which are detailed in Table 5.3, were set at levels that had previously been identified in the Plackett and Burman medium design trial experiments in Chapter 3 and also optimum L-cysteine.HCl.H$_2$O concentration achieved in chapter 4. The insignificant individual medium components identified by the Plackett and Burman medthod at low or high values relating to the greatest effects on total biogas volume were kept constant.

Table 5.3. Standard medium composition for the Method of Steepest Ascent medium design trails for both *C. beijerinckii* and *C. paraputrificum.*

Medium components	*C. beijerinckii*	*C. paraputrificum*
N-acetylglucosamine (GlcNAG)	X1	X1
Peptone	3 g/l	4 g/l
Yeast extract	2 g/l	2 g/l
Bufferig capacity	0.2M	0.2M
L-cysteine.HCl.H$_2$O	1 g/l	X2
FeSO$_4$.7H$_2$O	X2	0.4 g/l
MgCl$_2$	0.43 g/l	X3
Initial pH	X3	6.5

X1 GlcNAG (*C. beijerinckii*) = 12g/l, 17g/l, 22g/l, 27g/l, 32g/l
X1 GlcNAG (*C. paraputrificum*) = 12g/l, 17g/l, 22g/l, 27g/l, 32g/l, 37g/l
X2 FeSO$_4$.7H$_2$O = 0.1g/l, 0.08g/l, 0.06g/l, 0.04g/l, 0.02g/l
X3 Initial pH = 6.5, 6.3, 6.0, 5.7, 5.5

X2 L-cysteine.HCl.H$_2$O = 0.75g/l, 0.5g/l, 0.35g/l, 0.25g/l, 0.15g/l, 0.10g/l
X3 MgCl$_2$ = 0.3g/l, 0.35g/l, 0.4g/l, 0.45g/l, 0.5g/l, 0.55g/l

5.3 Results

5.3.1 Method of Steepest Ascent experimental results

The effect of changing medium component variables to achieve close to the optimum for total biogas production, utilizing N-acetylglucosamine as the sole carbon source, by *C. beijerinckii* and *C. paraputrificum* in anaerobic batch culture was investigated. Each fermentation experiment was run for 8 days, and the results were analysed, when the utilization of N-acetylglucosamine and production of biogas had ended. Production of biogas was assumed to have ended when there was no further noticeable increase in biogas volume observed for any treatment after 8 days. Tables 5.4 (*C. beijerinckii*) and table 5.5 (*C. paraputrificum*) show the correlation between the total volume of biogas produced and the medium composition trials.

Table 5.4 Experimental design and result of method of steepest ascent for *C. beijerinckii*.

Trial Number	N-acetylglucosamine (g/l)	FeSO$_4$.7H$_2$O (g/l)	Initial pH	Biogas Volume (ml/l)
1	12	0.10	6.5	4133
2	17	0.08	6.3	5733
3	22	0.06	6.0	4800
4	27	0.04	5.7	3233
5	32	0.02	5.5	666

g/l = grams per liter.
ml/l = milliliters of biogas per liter of culture medium.
n = 3

Table 5.5 Experimental design and result of method of steepest ascent for *C. paraputrificum*.

Trial Number	N-acetylglucosamine (g/l)	cysteine.HCl.H$_2$ (g/l)	MgCl$_2$ (g/l)	Biogas Volume (ml/l)
1	12	0.75	0.30	2366
2	17	0.50	0.35	3300
3	22	0.35	0.40	4433
4	27	0.25	0.45	5400
5	32	0.15	0.50	5333
6	37	0.10	0.55	2833

g/l = grams per liter.
ml/l = milliliters of biogas per liter of culture medium.
n = 3

The biogas yield from the *C. beijerinckii* cultures rose significantly ($P < 0.037$) to reach a peak in the second trial at concentrations N-acetylglucosamine 17 g/l, FeSO$_4$.7H$_2$O 0.08 g/l, initial pH of 6.3 before declining significantly ($P < 0.001$) in subsequent trials (Table 5.4). Consequently, the concentrations of medium constituents from trial 2 were chosen for the next phase of biogas production optimization. On the other hand, the biogas yield from the *C. paraputrificum* cultures rose significantly ($P < 0.001$) to reach a peak in the 4th trial run before declining significantly ($P < 0.001$) in subsequent trials (Table 5.5). Thus the concentrations of medium constituents from trial 4 were also chosen for the next phase of *C. paraputrificum* biogas production optimization.

The yield of biogas was observed to decline, after reaching its peak, when the medium constituents were further changed in the later trials with both *C. beijerinckii* and *C. paraputrificum*. Each of the experiments, to assess the optimum medium composition, lasted for a period of 8 days for both the *C.*

beijerinckii and the *C. paraputrificum* cultures.

Table 5.6 and table 5.7 show the concentration of organic acids produced compared to the amount of N-acetylglucosamine utilized for *C. beijerinckii* and *C. paraputrificum* respectively.

Table 5.6 Fermentation profile for *C. beijerinckii* when using N-acetylglucosamine as the main carbon source.

Trial Number	Biomass (g/l)			Initial GlcNAc (g/l)	GlcNAc Utilised (g/l)	Yield of product (g/l)		Biogas Volume (ml/l)
	Initial	Final	S.E			Acetate	Butyrate	
1	0.11	1.15	0.02	11	11.25	3.82	3.42	4133
2	0.09	1.98	0.01	16	15.80	4.51	4.59	5733
3	0.09	2.31	0.11	21	16.79	5.08	4.06	4800
4	0.11	1.61	0.01	27	10.39	3.27	2.81	3233
5	0.14	0.90	0.54	32	3.61	1.39	1.17	666

Trials 1 – 5 = Medium design trial experiments.
Biogas volume (ml/l) = Volume of biogas produced per liter of medium.
S.E. = Standard error of the mean
g/l = Biomass concentration in grams per liter.
n = 3

Table 5.7 Fermentation profile for *C. paraputrificum* when using N-acetylglucosamine as the main carbon source.

Trial Number	Biomass (g/l)			Initial GlcNAc (g/l)	GlcNAc Utilised (g/l)	Yield of product (g/l)			Biogas Volume (ml/l)
	Initial	Final	S.E			Lactate	Acetate	Butyrate	
1	0.20	0.42	0.01	12	11	0.90	3.92	1.93	2366
2	0.09	0.76	0.02	17	14	2.63	5.02	2.28	3300
3	0.07	0.96	0.02	22	19	4.14	6.38	2.88	4433
4	0.09	1.02	0.01	27	22	4.21	7.79	3.58	5400
5	0.14	1.45	0.09	32	25	5.02	8.07	3.69	5333
6	0.15	0.61	0.02	37	11	2.24	4.92	2.53	2833

Trials 1 – 5 = Medium design trial experiments
Biogas volume (ml/l) = Volume of biogas produced per liter of medium.
S.E = Standard error of the mean
g/l = Biomass concentration in grams per liter.
n = 3.

There were no significant links between the ratios of the organic acids produced to the volumes of cumulative biogas produced in C. beijerinckii and C. paraputrificum cultures. Analysis for   the correlations is shown in Appendix 5 (table 5.29 and 5.30). In each of the medium design trials for both *C. beijerinckii* and *C. paraputrificum*, the final pH was significantly lower than the initial pH ($P < 0.01$). For *C. beijerinckii*, the pH observed in trial 4 was significantly lower than in any of the other trials, at pH 5.04 whilst in trial 6 for *C. paraputrificum* the pH observed was significantly lower than other trials but the significant difference between between trial 5 and 6 was insignificant ($P > 0.05$). In addition the biomass concentration had also increased significantly after the 8 days of incubation from that initially inoculated ($P < 0.001$). For *C. beijerinckii*, the final biomass concentration observed in trial 3 was significantly higher than in any of the other trials, at 2.31 g/l. In contrast, for *C. beijerinckii*, the final biomass concentration observed in trial 5 was significantly lower than in any of the other trials, at 0.90 g/l.

Butyrate and acetate were the only organic acids detected from the *C. beijerinckii* medium design trial experiments. However acetate was generally the most abundant organic acid produced in all trial runs except in trial 2. However a student t-test revealed that there was no significant difference between the total concentrations of acetate and butyrate ($P > 0.05$). On the other hand, with the *C. paraputrificum* cultures lactate, acetate and butyrate were the only organic acids detected, and acetate was identified as the most abundant organic acid produced. Using a Post Hoc Tukey HSD multiple comparison test there were significant differences between organic acids produced within and between trials ($P < 0.05$) but overall, there was a significant difference between concentrations of acetate produced with lactate and butyrate ($P < 0.01$) but however no significant difference between butyrate and lactate (P > 0.05).

For *C. beijerinckii*, there was a significant total volume of biogas produced in all trials ($P < 0.01$). *C. beijerinckii* produced the greatest volume of biogas in trial 2 (table 5.5) at 5733 ml/l of biogas. Applying a multiple comparison using a Post Hoc Tukey HSD test there were significant differences between trial 2 and other trials ($P < 0.04$) except for trial 3 ($P > 0.05$). In other words total biogas volume produced in trials 1, 4 and 5 were significantly lower than trial 2. It was also noted that as the volume of biogas produced declined in trials 3, 4 and 5, so the total volume of organic acids produced also declined (Table 5.6). There was a significant correlation between the total volume of biogas produced and acetate ($P < 0.01$), butyrate ($P < 0.01$) and total concentrations of organic acids produced ($P < 0.01$). The smallest volume of biogas produced was recorded for medium design trial 5, in which a total of 666 ml/l of biogas was produced by the *C. beijerinckii* cultures. The total volume of biogas produced was significantly smaller compared to other trials ($P < 0.03$). In contrast the greatest volume of biogas was produced by *C. paraputrificum* were in medium design trial 4. Applying the one way ANOVA test there was a significant difference between the trials ($P < 0.001$), moreover Post Hoc Tukey HSD analysis reveal that the highest biogas produced were in trial 4 and 5, when compared with trial 1 ($P < 0.001$), trial 2 ($P < 0.001$), trial 3 ($P < 0.001$) and trial 6 ($P < 0.001$), however there was no significant difference between trial 4 and 5 ($P > 0.05$). The smallest volume of biogas (2833 ml/l) produced by *C. paraputrificum* was from the medium design trial 6. The total volume of biogas produced in trial 1 was also small and there were no significant difference between trial 1 and 6 (P > 0.05).

The amount of N-acetylglucosamine remaining in the *C. beijerinckii* medium of design trials 3, 4 and 5 was significantly greater than in trials 1 and 2, at 4.21 g/l, 16.6 g/l and 28.39 g/l after the 8 d of incubation (P < 0.05) (Table 5.6).

Table 5.8 and table 5.9 show the yield coefficient for metabolic products per unit of N-acetylglucosamine metabolized, in addition to the carbon recovery data for *C. beijerinckii* and *C. paraputrificum* respectively.

Table 5.8 Yield coefficients of metabolic products, and also carbon recovery for *C. beijerinckii*, using the Method of Steepest Ascent.

Trial Number	GlcNAc consumed (g/l)	Yield coefficient of product (mol of product/mol GlcNAc utilised)		Carbon recovery from GlcNAc (%)	Biogas Volume (ml/l)
		Acetate	Butyrate		
1	11.25	1.27	0.77	70	4133
2	15.80	1.07	0.74	64	5733
3	16.79	1.13	0.61	59	4800
4	10.39	1.18	0.69	64	3233
5	3.61	1.44	0.82	77	666

Trials 1 – 5 = Experimental trial designs
Biogas volume (ml/l) = Volume of biogas produced per liter of medium.
% = Percentages
g/l = Biomass concentration in grams per liter.
n = 3

Table 5.9 Yield coefficients of metabolic products, and also carbon recovery from the medium design trials, for *C. paraputrificum* using the Method of Steepest Ascent.

Trial Number	GlcNAc consumed (g/l)	Yield coefficient of product mol of product/mol of GlcNAc utilised			Carbon recovery from GlcNAc (%)	Biogas Volume (ml/l)
		Lactate	Acetate	Butyrate		
1	10.50	0.21	1.40	0.47	66	2366
2	14.20	0.46	1.32	0.41	71	3300
3	19.07	0.54	1.25	0.38	71	4433
4	22.45	0.47	1.30	0.40	70	5400
5	24.53	0.51	1.23	0.38	69	5333
6	12.09	0.46	1.52	0.53	82	2833

Trials 1 – 5 = Experimental trial designs
Biogas volume (ml/l) = Volume of biogas produced per liter of medium.
% = Percentages
g/l = Biomass concentration in grams per liter.
n = 3

Metabolite carbon recovery was calculated as the percentage of the number of moles of carbon contained in the organic acids produced by the clostridia per mole of N-acetylglucosamine metabolized. This was also calculated without considering the amount of carbon contained in cell biomass.

5.4 Discussion

The method of steepest ascent allows a series of rational changes to the medium to be tested until no further improvement in the product yield, in this case total volume of biogas produced, is achieved. The data from the experiments performed in this chapter indicate that the optimal conditions for biogas production by *C. beijerinckii* were 17 g/l N-acetylglucosamine, 0.08 g/l $FeSO_4.7H_2O$ and an initial pH 6.3, whilst for *C. paraputrificum* the optimal conditions were 27 g/l N-acetylglucosamine, 0.25 g/l L-cysteine.HCl.H_2O and 0.45 g/l $MgCl_2$ were close to the area of maximum total volume of biogas.

At the end of the fermentation period, when biogas and acid production had ceased, the medium pH was found to be at its lowest. Organic acids are responsible for pH decline and are also produced during bio-hydrogen production by Clostridia (Sim et al., 2007). Ginkel and Logan (2005) suggested that organic acids produced by the bacterial culture reduce bio-hydrogen yields.

Carbon source was limited in some trials due to exhaustion of initial N-acetylglucosamine supplied, this may also influence total volume of biogas produced. However, too high a level of carbon source can reduce microbial cell growth, it is necessary to be thorough when selecting an initial carbon concentration. A decrease in bacterial cell growth is usually detected when the concentration of primary carbon source is too high. At excessive concentration levels, dehydration of the bacteria cell occurs. The concentration of a carbon source is key during medium design to encourage the production of a desired secondary metabolite. When excessive carbon source concentrations are detrimental, repression of one or more enzymes needed for the synthesis of a secondary metabolite can occur (Greasham & Herber, 1997).

The *C. beijerinckii* cultures tested in medium design trial 5 produced the least biogas, and yet had been supplied with the greatest amount of N-acetylglucosamine. Consequently, it is possible that the excess N-acetylglucosamine supply led to catabolite repression or bacterial cell stress due to dehydration, since a substantial proportion of the N-acetylglucosamine remained (consumed 3.61 g/l out of an initial concentration of 32 g/l of N-acetylglucosamine) in the medium at the end of the experiment (Greasham & Herber, 1997). In contrast it was medium design trial 1, in which the *C. paraputrificum* cultures were supplied with limiting concentrations of N-acetylglucosamine that yielded the smallest volume of (Table 5.6). It is likely that the *C. paraputrificum* culture in these trials was suffering nutrient limitation, since all of the N-acetylglucosamine (11 g/l) supplied to the culture was consumed by it.

Previous studies of the thermodynamics for hydrogen production by *Enterobacter aerogenes* and attempts to optimize the medium for hydrogen production have concluded that the initial glucose concentration is important for the hydrogen yield during fermentation (Fabiano and Perego, 2002). These authors found that at relatively low initial glucose concentrations, the hydrogen production rate was also relatively low. From their experiments they deduced that over a threshold range of 20 to 30 g/l of glucose concentration, there was a sharp decline in hydrogen production. However the decline in total biogas produced for *C. paraputrificum* cultures started at 27 g/l of N-acetylglucosamine but for the *C. beijerinckii* cultures the decline in total biogas produced started at 17 g/l of N-acetylglucosamine. It is likely that at levels of N-acetylglucosamine that exceed the "threshold levels" observed for *C. beijerinckii* and *C. paraputrificum* then various metabolic pathways involved in fermentation may change or become inhibited (Fabiano & Perego, 2002). The results of the Method of Steepest Ascent experiments described in this chapter an N-acetylglucosamine concentration of 17 g/l was selected for the next phase of the experiments to optimize the culture medium for biogas production by *C. beijerinckii*, whilst a concentration of 27 g/l was chosen for *C. paraputrificum*.

Interestingly, once the optimum medium conditions for biogas production had been achieved, further increases in the medium components under test actually led to a significant decrease in biogas production. Increasing the amount of N-acetylglucosamine available for the cells of both species to metabolise led to a significant increase in biomass, until a point was reached beyond which further increases in N-acetylglucosamine availability led to a significant decrease in biomass. Previous studies have noted that as the concentration of the glucose supply for *C. beijerinckii* strain RZF-1108 was increased from 5 g/l to 9 g/l, the bio-hydrogen yield and production rate did indeed increase, but at concentrations of glucose above 9 g/l there was a decline in both bio-hydrogen yield and production rate (Zhao et al., 2011). Hydrogen production and bacteria cell concentration trends were the same as that of hydrogen yield and production rate. Whilst Lin et al. (2011) reported a maximum yield of 1891 ml H_2/l of medium from 10 g/l glucose, in the experiments reported in this work the maximum yield of biogas for *C. beijerinckii* cultures was 5733 ml biogas/l of medium (where biogas represents H_2 and CO_2 produced) from an initial N-acetylglucosamine concentration of 17 g/l. In common with the experiments using both *C. beijerinckii* and *C. paraputrificum* supplied with N-acetylglucosamine described in this work (Table 5.6 and Table 5.7), Zhao et al. (2011) noted that the glucose utilization by *C. beijerinckii* decreased, and so did the final pH, as the concentration of glucose supplied to the cultures was increased. It was also noted in that study that there was some residual glucose left over when the initial concentration was over 9 g/l, as there was residual N-acetylglucosamine in the current experiments when the initial concentration supplied exceeded the threshold values of 17 g/l and 27 g/l, for *C. beijerinckii* and *C. paraputrificum* respectively.

Clostridium cell growth and organic acid concentrations may have also decreased due to the effect of initial pH for *C. biejerinckii* cultures and also accumulation of organic acids for *C. beijerinckii* and *C. paraputrificum* cultures, which inhibited the volume of biogas produced (Table 5.6 and 5.7). It was noted that as the fermentation run carried on with different design trials there were different volumes of biogas produced; this helps to further emphasis the effect of the changing medium constituent variables.

It should be noted that the effect of the different variables assessed in a Method of Steepest Ascent analysis cannot be considered individually, since the variables are interdependent, however the effect of a variable may be more significant on an effect over the other variable. For the *C. beijerinckii* cultures production of a high total volume of biogas was significantly associated with an increase in the initial N-acetylglucosamine concentration at the same time as the initial concentration of $FeSO_4.7H_2O$ and culture medium pH were reduced. However once a certain threshold had been passed (demonstrated by Trial 2; Table 5.4), further increases in the N-acetylglucosamine concentration, along with reductions in the initial $FeSO_4.7H_2O$ concentration and culture medium pH resulted in a significant decrease in the total volume of biogas produced. The same effect was observed for the *C. paraputrificum* cultures in that a simultaneous increase in the initial concentrations of N-acetylglucosamine and $MgCl_2$ with a reduction in the initial concentration of L-cysteine.HCl.H_2O was significantly associated with an increase in the total volume of biogas produced, but again once a threshold had been passed (demonstrated by Trial 4; Table 5.5) there was a decrease in the total volume of biogas produced.

In Table 5.4, the medium design trial cultures of *C. biejerinckii* supplied with the highest concentration of N-acetylglucosamine (T5) showed a significantly reduced biogas yield (666 ml/l of medium); catabolite repression by N-acetylglucosamine cannot be totally responsible for this as the initial pH was also low (pH 5.5). At a very low pH of 4.0, *C. beijerinckii* RZF-1108 is known to produce no bio-hydrogen, however as the pH is raised from 5.0 to 7.0, so the bio-hydrogen yield increases until the pH is raised above 7.0, when it decreases again up to a maximum of pH 10 (Zhao et al., 2011). The *C. beijerinckii* RZF-1108 cell concentrations, cumulative bio-hydrogen yield and organic acid production showed a similar trend to hydrogen yield; the hydrogen production rate increased as the initial pH was increased, up to pH 7.0 (Zhao et al., 2011). Cultures of *C. biejerinckii* supplied with N-acetylglucosamine in the study described in this work showed a similar trend to those of Zhao et al. (2011) since the total volume of biogas produced, final cell density and amount of N-acetylglucosamine utilised by the cultures were also all dependent upon the initial pH.

A low initial pH in the growth medium, and exposure to low pH during during fermentation will usually lead to a low level of intracellular ATP, since ATP will be expended to excrete H^+ from the cell, which will results in the inhibition of bacterial growth (Bowles & Ellefson, 1985). During all of the medium design trials described in this work, there was a reduction in the pH of the medium; this was due to the production of organic acids concomitant with the metabolism of N-acetylglucosamine from the medium. Acetate and butyrate were produced by both *C. beijerinckii* and *C. paraputrificum* cultures during the experiments, however only *C. paraputrificum* produced lactate. Butyrate and acetate are known to be produced by Clostridia during bio-hydrogen fermentation (Sim et al., 2007). It is known that it is important to control the pH of a fermentation process accurately, since this can have an effect on the yield of products, due to its influence on bacterial metabolism.

Evvyernie et al. (2000) also found that the total volume of biogas produced was highly dependent upon the initial pH of the culture medium. Their research found that the a significantly larger volume of biogas was produced at an initial pH of 6.5, whilst at an initial pH in the medium of 6.0 or 7.0 there was a significant reduction in the volume of biogas evolved. During the Method of Steepest Ascent experiments with *C. beijerinckii* cultures described here, it was found that the optimum initial pH for total biogas produced was pH 6.3, which is very similar to the optimum recorded by Evvyernie et al. (2000). In addition, during the current work, different ratios and concentrations of organic acids were produced at the different pH ranges tested. Consequently, it seems

174

reasonable to assume that the fermentation products evolved varied depending on the initial pH of the medium as well. It has been observed previously that changes in the types of metabolic products formed can result from changes in the metabolic pathways used by microorganisms (Evvyernie et al., 2000). The production of hydrogen is concurrent with the production of organic acids as well as solvent production. During the statistical optimization of important variables for enhanced hydrogen production by *Clostridium tyrobutyricum*, pH of the culture decreased consistently with the production of hydrogen, formation of organic acids and ethanol. However the major metabolites produced that was associated with the evolution of hydrogen at pH 6.3 was butyrate (Ji Hye et al., 2008). The organic acids produced for *C. beijerinckii* cultures were acetate and butyrate but acetate was the most abundant organic acid produced.

During the Plackett and Burman medium design trials described in Chapter 3 $FeSO_4.7H_2O$ had a significant effect upon biogas production by *C. beijerinckii* when it was supplied at a low initial concentration (0.1 g/l). Further optimization of the medium using the Method of Steepest Ascent demonstrated that and $FeSO_4.7H_2O$ concentration of 0.08 g/l was associated with the production of a significantly greater volume of biogas. Since excess $FeSO_4.7H_2O$ in a medium can encourage precipitation of medium constituents, it is particularly important that $FeSO_4.7H_2O$ is present in only trace quantities. Previous workers have also noted that the $FeSO_4$ concentration plays a key role in the production of hydrogen at high cell densities by influencing the metabolic pathways of fermentative bacteria, with only small quantities of $FeSO_4$ sufficient significantly to improve hydrogen production (Lee et al., 2009). The effect of iron concentration during Clostridia fermentation on hydrogen production rate may be related not only to the iron concentration for the growth but also environmental conditions, which the presence of iron or iron concentration is affected (Lee et al., 2001).

The iron concentration in the growth medium has an effect on many different features of microbial cell physiology. For example the iron concentration to which *C. beijerinckii* RZF-1108 cultures were exposed stimulated hydrogen production, *hydA* gene expression and production of biomass (Zhao et al., 2012). Similarly, supplementation of the culture medium with Fe^{2+} during growth of anaerobic mixed culture improved hydrogen production and substrate utilization, increased hydrogenase activity, and also reduced the lag time (Yang and Shen, 2006). Since hydrogenases catalyze the reversible oxidation of molecular hydrogen (H_2), and the inclusion of excess $FeSO_4$ in the culture medium will suppress the activity of the hydrogenase enzymes (Lee et al., 2009). It is clear that Fe^{2+} concentrations, which enhance this, will be beneficial to hydrogen yield. In addition, sulphur salts, such as sulphide or sulphite, when present in a culture medium have also been found to suppress hydrogen production, as the hydrogenase enzyme cannot reduce the sulphate (Lee et al., 2009). Nevertheless, although excessive Fe^{2+} can have a negative impact upon hydrogenase enzyme activity, the presence of trace Fe^{2+} is still important as the hydrogenase activity in fermentative bacteria has been found to decrease when the Fe^{2+} is exhausted (Dabrock et al., 1992). The presence of both the iron and the sulphur from $FeSO_4$ is important in a medium to increase hydrogenase biosynthesis which will result in increased H_2 evolution (Lee et al., 2009). Therefore the concentration of $FeSO_4$ is an important criterion for biohydrogen production, and for the clostridia it seems that it must be present but at low concentrations. This may be the reason why the ideal concentrations of $FeSO_4.7H_2O$ from the Plackett and Burman method reduced from 0.1 g/l to 0.08 g/l in the Method of Steepest Ascent due to medium constituent demand by *C. beijerinckii*.

A variety of metal ions is required for the proper functioning of microbial cells. It is known that Mg^{2+} cations are required by bacterial cells, and that they have a number of intracellular functions, although these functions are not always well-understood (Hughes and Poole, 1989). Generally, Mg^{2+} cations are known to be associated with the stabilization of different biological systems such as cell walls, and protein shapes and structures, as well as acting as catalysts for different biochemical activities with the ability to trigger and control reactions. $MgCl_2$ is required by microbial cells in trace quantities (Greasham & Herber, 1997). Magnesium ions are needed in for enzyme activities and function to stabilize cell membranes, as well as being for the growth of vegetative cells as they help maintaining normal cell morphology (Zhao et al., 2012; Zhonghua et al., 1978). Previous workers have found that Mg^{2+} has a substantial role in both cell growth and bio-hydrogen production from glucose by *C. beijerinckii* RZF-1108 (Zhao et al., 2012). As has been observed for the Fe^{2+} concentration in the medium, bio-hydrogen production by *C. beijerinckii* RZF-1108 was found to increase with increasing Mg^{2+} concentration up to a critical value (0.4 g/l), after which it decreased (Zhao et al., 2012). Magnesium (as $MgSO_4.7H_2O$) has also been found to have a significant effect on bio-hydrogen production rate by *C. acetobutylicum* up to a critical concentration, above which the rate is significantly reduced (Alshiyab et al., 2008). The experiments performed in the current study also demonstrated that once the Mg^{2+} concentration in the medium had exceeded a critical level of 0.45 g/l $MgCl_2$ then there was a decline in the total biogas volume produced by *C. paraputrificum* (Table 5.5). Magnesium ions are essential for the proper metabolic functioning of bacteria since ATP is generally found in cells as a chelate of both ATP and magnesium ions (Zhao et al., 2012). In addition to its effect upon bio-hydrogen production, increases in the concentration of $MgCl_2$ in the medium led to increases in the final biomass concentrations achieved up to $MgCl_2$ concentration of 0.5 g/l and biomass decreased above a critical level of 0.5 g/l of $MgCl_2$. (Table 5.7). Both magnesium and iron are required in trace quantities, functioning as cofactors or activators of variety enzymatic systems and also as biosynthetic regulators of some secondary metabolites (Weinberg, 1989). Therefore it is vital for both biogas production and biomass yield that the Mg^{2+} provided in a bacterial growth medium design is at a suitable concentration.

The total volume of biogas produced increased as the initial L-cysteine.HCl.H_2O concentration in the medium was reduced; however below a critical level of 0.25 g/l of L-cysteine.HCl.H2O the total volume of biogas reduced (Table 5.7). Previous workers have also found that the total hydrogen yield, hydrogen production rate and biomass yield all decreased as the L-cysteine concentration in the medium was increased (Zhao et al., 2012). L-cysteine is a reducing agent that is often used to reduce the oxidation-reduction potential of a growth medium (Yang & Shen, 2006). It is difficult for strict anaerobes to survive when exposed to oxygen because they lack the enzymes catalase and superoxide dismutase, which help to convert to harmless water the harmful hydrogen peroxide and superoxide radicals formed in cells when they are exposed to oxygen. Clostridia are strict anaerobes and so they are not able to initiate growth except when the culture medium redox potential is controlled at an appropriately low level. Consequently, it is necessary to eliminate the dissolved oxygen from the growth medium to avoid oxygen toxicity, and cysteine is one of the medium additives that can be used to assist this. Previously in this work, in the one factor cysteine experiments, it was found that cultures grown in the presence of high cysteine concentrations produced significantly less biogas and biomass as compared with those cultivated in low concentrations of cysteine (Chapter 4, Table 4.1). The same was the case in table 5.6 as L-cysteine.HCl.H_2O concentrations decrease to 0.15 g/l, the total biogas concentrations increase and the final cell concentrations increase. Both of which indicate that it is essential to incorporate L-cysteine.HCl.H_2O in the growth medium, but that the concentration needs to be selected carefully to optimize the yield of the desired product (Table 5.7).

In the Method of Steepest Ascent medium design trials the medium constituent variables were varied simultaneously, using the results from the Plackett and

Burman analysis as the starting point. By using the Method of Steepest Ascent, total biogas production was further increased over the Plackett and Burman method from 3183 ml/l to 5733 ml/l of biogas for *C. beijerinckii* and from 1300 ml/l to 5400 ml/l of biogas for the *C. paraputrificum* cultures. However, the results of the Method of Steepest Ascent medium design trials cannot be taken individually, since the variables are interdependent and biogas production is not dependent on a single variable but certain variable effect are more dominant and significant over the other. The Box-Behnken method for rational medium design, as described in chapter 6, can be used to determine the interactions between individual medium constituents and also the optimum conditions for those medium constituents that will maximise biogas yield.

5.5 Conclusion

For *C. beijerinckii*, using the Method of Steepest Ascent, the optimum initial concentrations of N-acetylglucosamine and $FeSO4.7H_2O$ were determined to be 17 g/l and 0.08 g/l, respectively, whilst an initial pH of 6.3 was the optimum, yielding a maximum total volume of biogas of 5733 ml/l. On the other hand, using the Method Steepest Ascent, for *C. paraputrificum* the optimum N-acetylglucosamine, $L-cysteine.HCl.H_2O$ and $MgCl_2$ concentrations were found to be 27 g/l 0.25 g/l and 0.45 g/l, respectively, yielding a maximum total biogas volume of 5400 ml/l.

This technique shows an improvement as the total biogas produced in the Plackett and Burman method were further increased at the end of the Method of

Steepest Ascent experiment for *C. beijerinckii* and *C. paraputrificum*. However the Method of Steepest Ascent does not explain the interactions between variables involved in the fermentation process. A further optimization step is required to explain the interactions between variables; this was done using the Box-Behnken method and is described in Chapter 6.

Chapter 6

Optimisation of *Clostridium beijerinckii* and *C. paraputrificum* growth Medium for Biogas Production Using the Box-Behnken Method

6.1.1 Background

When investigating the effect of different medium constituent variables on a response in a controlled experiment, the first difficulty is to identify the independent medium constituent variables that influence the desired product. After these independent variables have been identified, the next step is to study their effects on a dependent response. The Box-Behnken design is an independent quadratic design used to study the quadratic effect of medium constituent variables, after the significant variables have been identified, by screening a series of factorial experiments. An example of a screening method that can be used to identify the significant variables is the Plackett and Burman method, which was used previously to identify the significant constituents of the growth medium (Chapter 3).

Each run of the fermentation experiment using the Box-Behnken method involves a combination of the factor levels that are to be investigated. The Box-Behnken method requires only 3 levels to run an experiment and tackles the problem of where the experimental boundaries should be in order to avoid experimental trial runs that are extreme in the design of experiment. This method enables each factor or variable level to be tested several times. Each independent medium constituent variable can be used at one of three equally spaced pre-determined values; this can be explained by using the coded values, which are -1, 0, and +1. This design also postulates an excellent predictability within the spherical design space. It has the ability to predict a desired response result and requires less experimental work and fewer experimental runs as compared with the full factorial design or central composite designs (Manikandan et al., 2010).

The Box-Behnken method has been applied in previous studies to develop the optimum medium formulation for bio-hydrogen production. Using *Enterobacter aerogenes* MTCC 111, Karthic et al. (2012) assessed the individual and interactive influences of significant process factors to optimize the production of biohydrogen from glucose as a substrate. A number of significant factors (glucose, initial pH and ferric chloride) were identified by these authors and then further optimized using the Box-Behnken method. Using the Box-Behnken method a maximum yield of 1.69 mol H_2/mol glucose was achieved from a medium with an optimum glucose concentration of 16.56 g/l, an initial pH of 6.15 and a ferric chloride concentration of 213.13 mg/l.

In the work described in this work, the optimum medium formulation for the production of biogas by *Clostridium beijerinckii* and *Clostridium paraputrificum* from utilizing N-acetylglucosamine was determined using the Box-Behnken design.

6.1.2 Objectives

- To determine the optimum levels of the various medium constituents required for the maximum total volume of biogas to be produced by *Clostridium beijerinckii* and *Clostridium paraputrificum*.
- To study the interactive effects of the medium constituents tested.

6.2 Materials and Methods

The fermentation and analytical techniques for the Box-Behnken design experiments were all carried out aseptically and in triplicate. The Clostridium strains, their storage conditions and their revival for the experiments described in this chapter were as described in Section 2.1.1. The starter cultures used as the initial inoculum for the experiments in this chapter were prepared as described in Section 2.1.5. Cell biomass analysis and biomass analysis were as described in 2.3.2. Metabolic products were prepared and analyzed as described in section 2.4.

For the 1-litre bioreactor batch experiments, pH measurements made using a bench pH meter were used to verify the pH within the bioreactor vessel. The optimized medium for the batch bioreactor fermentation run was prepared according to the results of the statistically-designed Box-Behnken experiments in this chapter.

The final biogas mixture was assumed to be H_2 and CO_2 produced form the headspace of the culture medium in syringes and Bioreactor run. Without considering the carbon contained in cell biomass the carbon recovery was calculated as the carbon contained in the organic metabolites only. The ideal gas conditions were applied to derive the concentration of each biogas produced from the total mixture of biogas and was calculated assuming the conditions were temperature of 37°C and pressure of 1 atmosphere. The difference between the sum of hydrogen contained in the number or moles of organic acids per N-acetylglucosamine utilised and the number of moles of hydrogen contained per mole of N-acetylglucosamine was assumed to be the theoretical hydrogen produced. This was also calculated as the mole of hydrogen gas derived per mole of N-acetylglucosamine utilized. (Appendix 4.2).

The Minitab software was used to construct the response surface curves and its corresponding contour curves described by the regression model. Using the Minitab software regression analysis was the statistical technique used for estimating the relationships among variables.

6.2.1 Experimental Design

Clostridium beijerinckii and *C. paraputrificum* were both cultured in 50 ml syringes to capture the biogas evolved. All experiments were carried out independently in triplicate. The optimum medium formulation was achieved by conducting 15 experiments in triplicate in order to find the combinations of the variables involved that would achieve the maximum possible total volume of biogas produced. The working values for the variables N-acetylglucosamine concentration, $FeSO_4.7H_2O$ concentration and Initial pH to optimize the medium for *C. beijerinckii* fermentations are shown in Table 6.1.

Table 6.1 The levels of the different variables used for the Box-Behnken design medium optimization experiments for *C. beijerinckii*.

Coded Values	N-acetylglucosamine (g/l)	FeSO4.7H2O (g/l)	Initial pH
-1	13	0.06	6.1
0	17	0.08	6.3
1	21	0.10	6.5

-1 = low levels
0 = medium levels

186

1 = high levels.
g/l = grams per litre

The working values for the variables N-acetylglucosamine concentration, L-cysteine.HCl.H$_2$O concentration and MgCl$_2$ concentration to optimize the medium for *C. paraputrificum* fermentations are shown in Table 6.2.

Table 6.2 The levels of the different variables used for the Box-Behnken design medium optimization experiments for *C. paraputrificum*.

Coded Values	N-acetylglucosamine (g/l)	L-cyesteine.HCl (g/l)	MgCl$_2$ (g/l)
-1	23	0.10	0.40
0	27	0.25	0.45
1	31	0.40	0.50

-1 = low levels
0 = medium levels
1 = high levels.
g/l = grams per litre

The total volumes of biogas produced were influenced by the simultaneous changes in trial levels of these medium constituent variables used in the fermentation medium. Table 6.3 shows the levels of the variables used in the various medium design trials for the Box-Behnken design experiments performed in the current work.

Table 6.3 The levels of the medium constituents used for culturing *C. beijerinckii* and *C. paraputrificum* in the Box-Behnken design experiments.

Trial	Independent variable codes		
Number	X_1	X_2	X_3
1	-1	-1	0
2	1	-1	0
3	-1	1	0
4	1	1	0
5	-1	0	-1
6	1	0	-1
7	-1	0	1
8	1	0	1
9	0	-1	-1
10	0	1	-1
11	0	-1	1
12	0	1	1
13	0	0	0
14	0	0	0
15	0	0	0

X1 to X3 = Independent medium variables, as shown in Tables 6.1 and 6.2

In medium design trials 1–12 the clostridia were cultured with different medium constituent combinations, whilst in trials 13–15 they were cultured with unchanged medium constituent concentrations. N-acetylglucosamine, FeSO$_4$.7H$_2$O and initial pH were the three variables tested to optimise *C. beijerinckii* fermentation, whilst for *C. paraputrificum* they the variables tested were N-acetylglucosamine, L-cysteine.HCl.H$_2$O and MgCl$_2$.

A total of 15 different medium formulations were tested by changing each of the three independent variables, using the Box–Behnken design. The centre point (ie constituent levels 0, 0 and 0) was replicated three times to permit the level of experimental error to be calculated. The results of the 15 Box-Behnken medium design trials were then used to determine the optimum values for each of the medium constituents using Response surface methodology. In order to predict the optimum values for each of the medium constituents to achieve the desired response, the quadratic polynomial equation was applied to connect the relationship between the independent variables and the desired response (*i.e.* total biogas volume). The statistical software package Minitab version 16.1.0, (Minitab Inc, Pennsylvania, USA) was used to analyze the results using variance effects, standard errors and Student's T-test.

6.2.2 Maximum desirability for Optimum Medium Constituent.

Based on the results of the Box-Behnken medium design trials described above, the optimized media for biogas production by *C. beijerinckii* and *C. paraputrificum* were prepared to the formulation descibed in Table 6.4. The medium formulations described in Table 6.4 were used for the syringe and 1 litre bioreactor batch fermentation in this chapter.

Table 6.4 Box-Behnken optimized media for biogas production by *C. beijerinckii* and *C. paraputrificum*.

Variable Code	Medium Variables	*C. beijerinckii*	*C. paraputrificum*
		\multicolumn{2}{	}{Variable concentrations}
X1	N-acetylglucosamine	21 g/l	29.38 g/l
X2	Peptone	3 g/l	4 g/l
X3	Yeast extract	2 g/l	2 g/l
X4	Buffer capacity	0.2 M	0.2 M
X5	L-cysteine.HCl	1 g/l	0.27 g/l
X6	$FeSO_4.7H_2O$	0.1 g/l	0.4 g/l
X7	$MgCl_2$	0.427 g/l	0.40 g/l
X8	Initial pH	6.11	6.5

g/l = grams per litre.

6.3 Results

6.3.1 Analysis of Box-Behnken medium design trials for *C. beijerinckii* and *C. paraputrificum.*

The effects of interactions between the various medium constituents and also the optimum medium formulation or maximum total biogas production from N-acetylglucosamine by *C. beijerinckii* and *C. paraputrificum* were investigated. The organic acids detected in the *C. beijerinckii* medium at the end of the fermentation were butyrate and acetate, with acetate being significantly more abundant than butyrate ($P < 0.001$), (Table 6.5). On the other hand, the organic acids detected in the *C. paraputrificum* medium at the end of the fermentation were lactate, formate, acetate and butyrate. Ethanol was also produced by the *C. paraputrificum* cultures, but at lower concentrations than the other metabolites, but there was no significant difference between ethanol and formate concentrations produced ($P > 0.05$). Acetate was significantly more abundant than any of the other organic acids produced ($P < 0.001$) followed by butyrate, lactate and then formate (Table 6.6). The acid concentrations in the spent culture medium were determined at the end of each fermentation experiment, when the utilization of N-acetylglucosamine had ceased and there were no further noticeable production of biogas.

Through the course of each fermentation experiment, the final pH was significantly lower than the initial pH ($P < 0.001$). There was also a substantial reduction in the residual N-acetylglucosamine concentration in the spent culture medium from both the *C. beijerinckii* and the *C. paraputrificum* fermentations

(Tables 6.5 and 6.6). The fermentation run for investigating the Box-Behnken design on biogas production lasted for duration of 18 days. Tables 6.5 and 6.6 summarise the fermentation profiles for *C. beijerinckii* and *C. paraputrificum* respectively.

Table 6.5. The N-acetylglucosamine fermentation profiles for *C. beijerinckii* in the Box-Behnkendesign experiments.

Trial	pH		Final biomass	Initial	GlcNAc	Yield of product (g/l)		Biogas
Number	Initial	Final	(g/l)	GlcNAc (g/l)	Utilised (g/l)	Acetate	Butyrate	Volume (ml/l)
1	6.3	5.5	1.91	13	12.22	4.11	3.61	4567
2	6.3	5.5	2.20	21	19.31	5.73	4.73	6367
3	6.3	5.5	1.60	13	12.58	3.70	3.90	4200
4	6.3	5.5	2.08	21	19.72	5.60	4.72	6300
5	6.1	5.3	1.92	13	12.38	3.62	3.58	3867
6	6.1	5.3	2.01	21	17.52	5.00	4.65	6067
7	6.5	5.8	1.17	13	12.87	4.14	3.62	3750
8	6.5	5.7	1.68	21	17.36	5.33	4.75	5300
9	6.1	5.3	1.98	17	16.52	4.50	4.69	4933
10	6.1	5.3	2.02	17	16.33	3.50	3.20	5200
11	6.5	5.7	1.47	17	14.73	5.00	4.16	5800
12	6.5	5.7	1.80	17	16.21	4.97	4.38	5100
13	6.3	5.4	1.77	17	16.33	4.92	4.21	5067
14	6.3	5.4	2.02	17	16.18	4.79	4.24	5100
15	6.3	5.5	2.01	17	17.08	5.02	4.34	5100

g/l = grams per litre.
ml/l = millilitres per litre.
n = 3

191

Table 6.6. The N-acetylglucosamine fermentation profiles for *C. paraputrificum* in the Box-Behnkendesign experiments.

Trial Number	pH Initial	pH Final	Final biomass (g/l)	Initial GlcNAc (g/l)	GlcNAc Utilised (g/l)	Yield of product (g/l) Lactate	Formate	Acetate	Ethanol	Butyrate	Biogas Volume (ml/l)
1	6.5	5.10	1.31	23	20.16	2.52	0.00	5.85	0.00	3.00	4100
2	6.5	4.99	1.41	31	25.62	3.62	0.41	6.58	0.44	2.45	4700
3	6.5	5.20	1.29	23	18.16	2.35	0.38	5.16	0.00	2.08	3433
4	6.5	5.03	1.12	31	24.74	3.19	0.39	6.27	0.56	2.50	4900
5	6.5	5.17	0.82	23	19.64	1.57	0.00	5.68	0.00	3.19	4700
6	6.5	5.02	1.25	31	26.31	2.71	0.31	6.48	0.00	3.21	5166
7	6.5	5.10	1.17	23	19.34	3.23	0.41	5.89	0.00	2.17	4300
8	6.5	4.99	1.34	31	25.61	3.03	0.34	6.53	0.00	3.05	4866
9	6.5	5.11	1.45	27	22.68	2.10	0.20	6.18	0.00	3.38	5166
10	6.5	5.12	0.99	27	22.57	2.14	0.20	5.84	0.00	3.00	4933
11	6.5	5.12	1.21	27	23.00	2.26	0.29	6.01	0.00	3.29	4966
12	6.5	5.18	0.99	27	22.79	2.02	0.00	5.31	0.00	2.94	4200
13	6.5	5.08	1.13	27	22.17	3.13	0.33	5.68	0.00	2.54	4900
14	6.5	5.11	1.23	27	22.70	2.21	0.31	5.97	0.00	3.12	4900
15	6.5	5.11	1.19	27	22.99	2.19	0.31	6.01	0.00	3.30	4866

g/l = grams per litre.
ml/l = millilitres per litre.
n = 3

There was an increase in biomass concentration in all-experimental trials. Organic acids were produced in all the trials. Trial 2 for the *C. beijerinckii* cultures (table 6.5) had the highest volume of biogas produced (6367 ml/l) whilst trial 6 and trial 9 (table 6.6) for the *C. paraputrificum* cultures had the highest total volume of biogas produced (5166 ml/l). The lowest volume of biogas produced *C. beijerinckii* was recorded in the 7th experimental trial and a total of 3750 ml/l of biogas was produced (table 6.5) while for *C. paraputrificum* the lowest total biogas volume of 3433 ml/l was produced in trial 3 (table 6.6).

Table 6.7 (*C.beijerinckii*) and table 6.8 (*C. paraputrificum*) shows the yield coefficient of the metabolized products per utilized N-acetylglucosamine as well as the recovery of carbon throughout the cultivation.

Table 6.7. Metabolic product yield coefficients, and carbon recovery, for cultures of *C. beijerinckii* in the Box-Behnken medium design trials.

Trial Number	GlcNAc Utilised (g/l)	Yield coefficient of product (mol of product/mol GlcNAc utilised)		Carbon recovery from GlcNAc (%)	Biogas Volume (ml/l)
		Acetate	Butyrate		
1	12.22	1.26	0.75	69	4567
2	19.31	1.11	0.62	59	6367
3	12.58	1.10	0.79	67	4200
4	19.72	1.06	0.61	57	6300
5	12.38	1.10	0.73	64	3867
6	17.52	1.07	0.67	60	6067
7	12.87	1.21	0.71	66	3750
8	17.36	1.15	0.69	63	5300
9	16.52	1.02	0.72	62	4933
10	16.33	0.80	0.50	45	5200
11	14.73	1.27	0.72	68	5800
12	16.21	1.15	0.69	63	5100
13	16.33	1.13	0.66	61	5067
14	16.18	1.11	0.67	61	5100
15	17.08	1.10	0.65	60	5100

g/l = grams per litre.
ml/l = millilitres per litre.
% = Percentage.
n = 3

Table 6.8. Metabolic product yield coefficients, and carbon recovery, for cultures of *C. paraputrificum* in the Box-Behnken medium design trials.

Trials Number	GlcNAc Consumed (g/l)	Yield coefficient of product (mol of product/mol GlcNAc utilised)					Carbon recovery from GlcNAc %	Biogas Volume (ml/l)
		Lactate	Formate	Acetate	Ethanol	Butyrate		
1	20.16	0.31	0.00	1.09	0.00	0.38	57.70	4100
2	25.62	0.35	0.08	0.96	0.16	0.24	54.41	4700
3	18.16	0.32	0.10	1.06	0.00	0.29	54.49	3433
4	24.74	0.32	0.08	0.95	0.22	0.26	55.00	4900
5	19.64	0.20	0.00	1.08	0.00	0.41	55.13	4700
6	26.31	0.26	0.06	0.92	0.00	0.31	48.86	5166
7	19.34	0.41	0.11	1.14	0.00	0.29	59.63	4300
8	25.61	0.29	0.06	0.96	0.00	0.30	50.82	4866
9	22.68	0.23	0.04	1.02	0.00	0.38	53.63	5166
10	22.57	0.24	0.04	0.97	0.00	0.34	50.50	4933
11	23.00	0.24	0.06	0.98	0.00	0.36	52.61	4966
12	22.79	0.22	0.00	0.87	0.00	0.33	46.44	4200
13	22.17	0.35	0.07	0.96	0.00	0.29	52.63	4900
14	22.70	0.24	0.07	0.98	0.00	0.35	52.02	4900
15	22.99	0.24	0.07	0.98	0.00	0.36	52.40	4866

g/l = grams per litre.

ml/l = millilitres per litre.
% = Percentage.
$n = 3$

The initial N-acetylglucosamine concentration in the growth medium, the cell biomass concentration, the initial pH and the total biogas volume all changed according to the changes in each Box-Behnken growth medium trial designs. The carbon recovery in the experiments was calculated as the percentages of the total amount of carbon, in moles, contained in acetate, butyrate, formate, lactate and ethanol produced per mole of N-acetylglucosamine utilized without considering the amount of carbon in cell biomass.

Employing multiple regression analysis on the fermentation experimental results recorded, the resulting second-order polynomial equation was establish to describe the total biogas production by the two clostridial species tested:

Clostridium beijerinckii

$$Y = 5088.67 + 956.25 X_1 - 108.12 X_2 - 14.37 X_3 - 121.71 X_1^2 + 391.04 X_2^2 - 221.46 X_3^2 + 75 X_1X_2 - 162.5 X_1X_3 - 241.75 X_2X_3 \text{ .... Equation 6.1}$$

Clostridium paraputrificum

$$Y = 4888.67 + 387.38 X_1 - 183.25 X_2 - 204.12 X_3 - 331.83 X_1^2 - 273.58 X_2^2 + 201.17 X_3^2 + 216.75 X_1X_2 + 25 X_1X_3 - 133.25 X_2X_3 \text{ ......Equation 6.2}$$

Symbol **Y** is the total volume of biogas predicted to be produced. For the *C. beijerinckii* cultures symbols X_1, X_2, and X_3 refer to the initial levels of N-acetylglucosamine and $FeSO_4.7H_2O$ in the medium, and also the initial pH, respectively. For the *C. paraputificum* cultures X_1, X_2, and X_3 refer to the initial levels of N-acetylglucosamine, L-cysteine.HCl.H_2O and $MgCl_2$, respectively, in the medium. The analysis of estimated regression coefficient for biogas was undertaken using coded units and used to evaluate the individual effects. The regression analyses comparing the impact of the different medium constituents tested with the clostridial cultures are shown in Tables 6.9 and 6.10.

Table 6.9. Regression analysis of the effect of the various medium constituents tested upon total biogas volume produced by *C. beijerinckii* in the Box-Behnken medium design trials

Term	Coefficient	Standard error	T- Value	P-Value
Constant	5088.67	151.32	33.628	0.000
X_1	956.25	92.67	10.319	0.000
X_2	-108.12	92.67	-1.167	0.296
X_3	-14.37	92.67	-0.155	0.883
X_1^2	-121.71	136.40	-0.892	0.413
X_2^2	391.04	136.40	2.867	0.035
X_3^2	-221.46	136.40	-1.624	0.165
$X_1 X_2$	75.00	131.05	0.572	0.592
$X_1 X_3$	-162.50	131.05	-1.240	0.270
$X_2 X_3$	241.75	131.05	-1.845	0.124

$X_1 - X_3$ = Independent medium constituent variables.
X_1 = N-acetylglucosamine.
X_2 = $FeSO_4.7H_2O$.
X_3 = Initial pH.
n = 3

Table 6.10. Regression analysis of the effect of the various medium constituents tested upon total biogas volume produced by *C. paraputrificum* in the Box-Behnken medium design trials.

Term	Coefficient	Standard error	T- Value	P-Value
Constant	4888.67	108.58	45.024	0.000
X_1	387.38	66.49	5.826	0.002
X_2	-183.25	66.49	-2.756	0.040
X_3	-204.12	66.49	-3.070	0.028
X_1^2	-331.83	97.87	-3.391	0.019
X_2^2	-273.58	97.87	-2.795	0.038
X_3^2	201.17	97.87	2.055	0.095
$X_1 X_2$	216.75	94.03	2.305	0.069
$X_1 X_3$	25.00	94.03	0.266	0.801
$X_2 X_3$	-133.25	94.03	-1.417	0.216

$X_1 - X_3$ = Independent medium constituent variables.
X_1 = N-acetylglucosamine.
X_2 = L-cysteine.HCl.H_2O.
X_3 = $MgCl_2$.
n = 3

For the *C. beijerinckii* cultures the variable estimate and the resultant p-values in table 6.9 imply that the linear term of effect of X_1 (N-acetylglucosamine) with p-value of 0.0 is significant on total biogas production. The model contains 3 squared effects (X_1^2, X_2^2, X_3^2) each for *C. beijerinckii* and *C. paraputrificum* cultures. The square term X_2^2 in the *C. beijerinckii* culture with p-value of 0.035 are significant on total biogas production. The square term of $FeSO_4.7H_2O$ had significant effect on the correlation of coefficients associated with low p-values of less than 0.05. Linear terms X_1, X_2, X_3 and squared term X_1^2 and X_2^2 were significant with individual p-values less than 0.05 for the *C. paraputrificum* cultures (Table 6.10). From the analysis of variance result, the test whether the terms in the model have effects on the total biogas produced were analyzed and the regression model was identified to be significant ($P < 0.005$) for *C. beijerinckii* and for *C. paraputrificum* (P < 0.013) indicating a high significance of the model. The R^2 values were 0.9618 (*C. beijerinckii*) and 0.9429 (*C. paraputrificum*).

Using the coded values, from equations resulting from differentiation of equation 6.1 and 6.2, the optimum values of X_1, X_2 and X_3 were evaluated to be 1, 1 and -0.9394 for *C. beijerinckii* and 0.5960, 0.1515 and -1 for *C. paraputrificum* respectively. The optimum values of the model for *C. beijerinckii* were 21 g/l of N-acetylglucosamine, 0.1 g/l of $FeSO_4.7H_2O$ and initial pH of 6.11 and for *C. paraputrificum* were 29 g/l of N-acetylglucosamine, 0.27 g/l of L-cysteine.$HCl.H_2O$ and 0.4 g/l of $MgCl_2$. The maximum predicted values for total biogas volume were 6478.95 ml/l (*C. beijerinckii*) and 5397.78 ml/l (*C. paraputrificum*).

The model contains 3 two-way interactions X_1X_2, X_1X_3 and X_2X_3 each for *C. beijerinckii* and *C. paraputrificum*, because there are 3 interaction terms as shown in tables 6.9 and 6.10. The results were analysed using response surface plots and the contour plots resulting from these. These plots help to clarify the interactions of the independent variables. Contour plots are drawn in two dimensions and they represent the relationship between the response (volume of total biogas) and the experimental variables under consideration (eg initial N-acetylglucosamine and iron concentrations). Points with the same response are linked to create contours of constant responses. In this analysis these plots show the way the response relates to two variables, according to the model equation, whilst the other variable is kept constant. Each response surface plot shows the effect of changing two independent variables whilst maintaining the third independent variable at a fixed level. The shapes of the resulting contour plots, which can be elliptical or circular, will show whether the interactions between the independent variables are significant or not (Pan et al., 2008). The contours represent a constant response (the total volume of biogas produced) to a variable stimulus. In the plots presented in the subsequent figures, the shade of green indicates the biogas yield, with the areas of darkest green colour representing the vicinity of highest total volume of biogas produced.

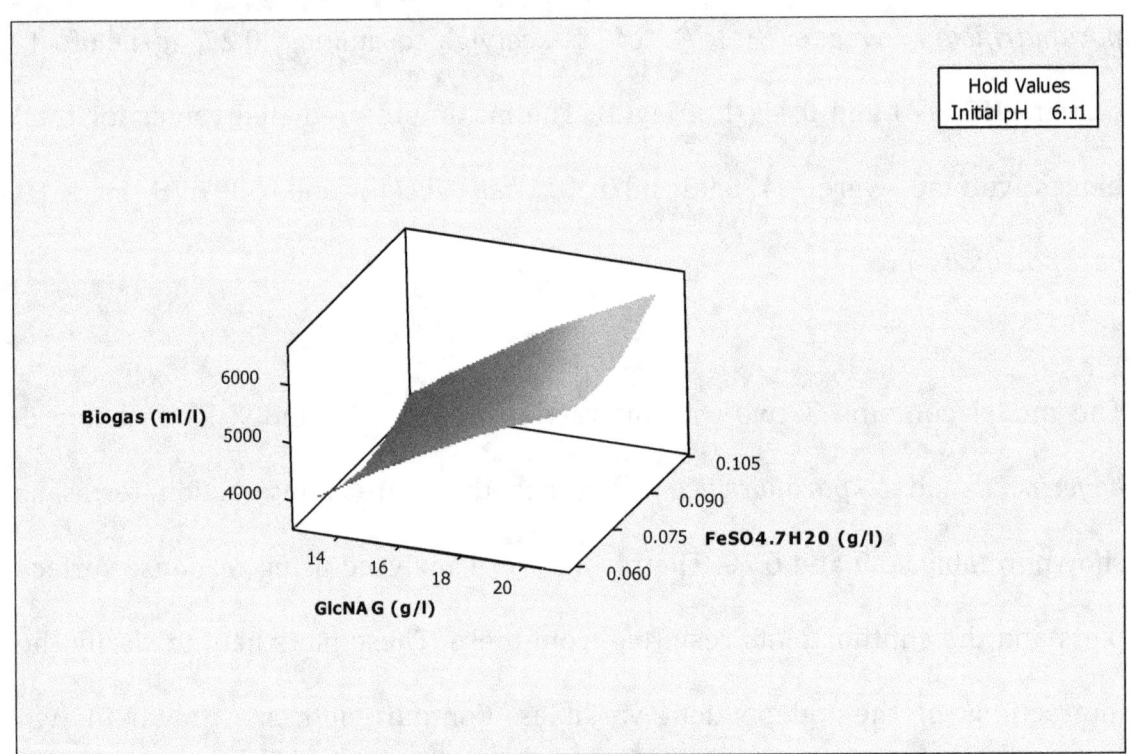

Figure 6.1 *Clostridium beijerinckii* biogas production response surface plot showing the effect of N-acetylglucosamine and FeSO$_4$.7H$_2$O with optimum level of initial pH of 6.11.

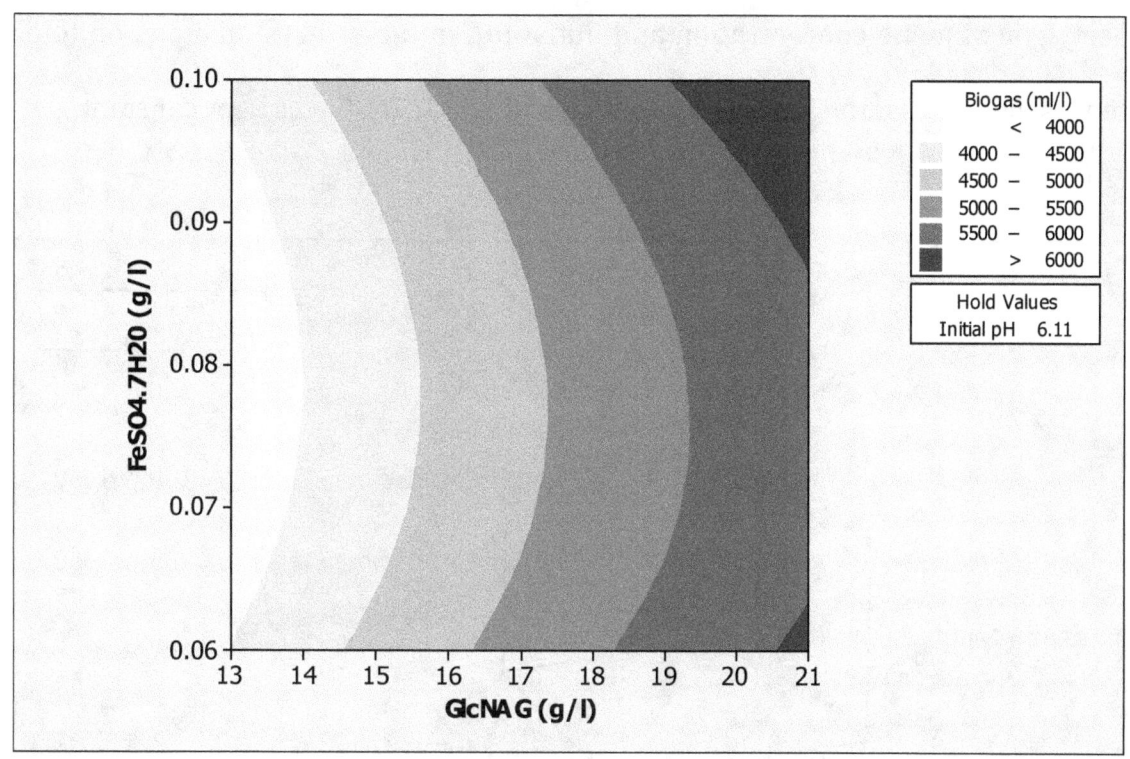

Figure 6.2 *Clostridium beijerinckii* biogas production contour plots showing the effect of N-acetylglucosamine and $FeSO_4.7H_2O$ with optimum level of initial pH 6.11.

Figure 6.1 and 6.2 show the response surface plot and resulting contour plots for independent variables N-acetylglucosamine (X1) and $FeSO_4.7H_2O$ (X2) while the third variable initial pH (X3) was constant at its optimal level. There was a substantial increase in total biogas volume when the concentration of N-acetylglucosamine increased. There was a slight decline in total biogas volume as $FeSO_4.7H_2O$ concentration increased from 0.06 to 0.075 g/l. The total volume of biogas increased slightly as the concentration of $FeSO_4.7H_2O$ increased from 0.08 to 0.10 g/l. However the positive effect of N-acetylglucosamine concentration on total volume of biogas produced was more notable. There was also a slight interaction effect among the variables of N-acetylglucosamine and $FeSO_4.7H_2O$ but the interaction effect was not significant ($P > 0.05$). Figures 6.3 and 6.4 show the interaction effects of the independent variables of initial N-acetylglucosamine concentration and initial pH, in the growth medium, on total

biogas yield whilst the initial concentration of FeSO$_4$.7H$_2$0 was kept constant.

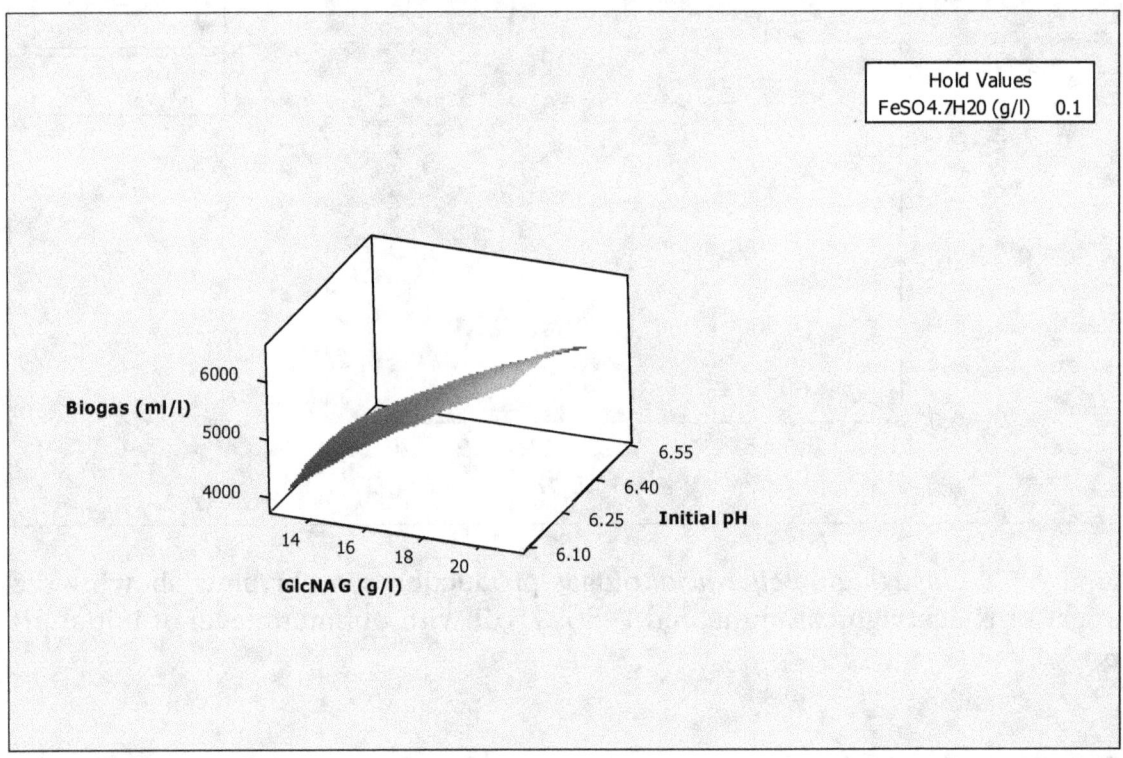

Figure 6.3. *Clostridium beijerinckii* biogas production response surface plots showing the effect of initial N-acetylglucosamine concentrations and initial pH with optimum level of FeSO$_4$.7H$_2$O of 0.1 g/l.

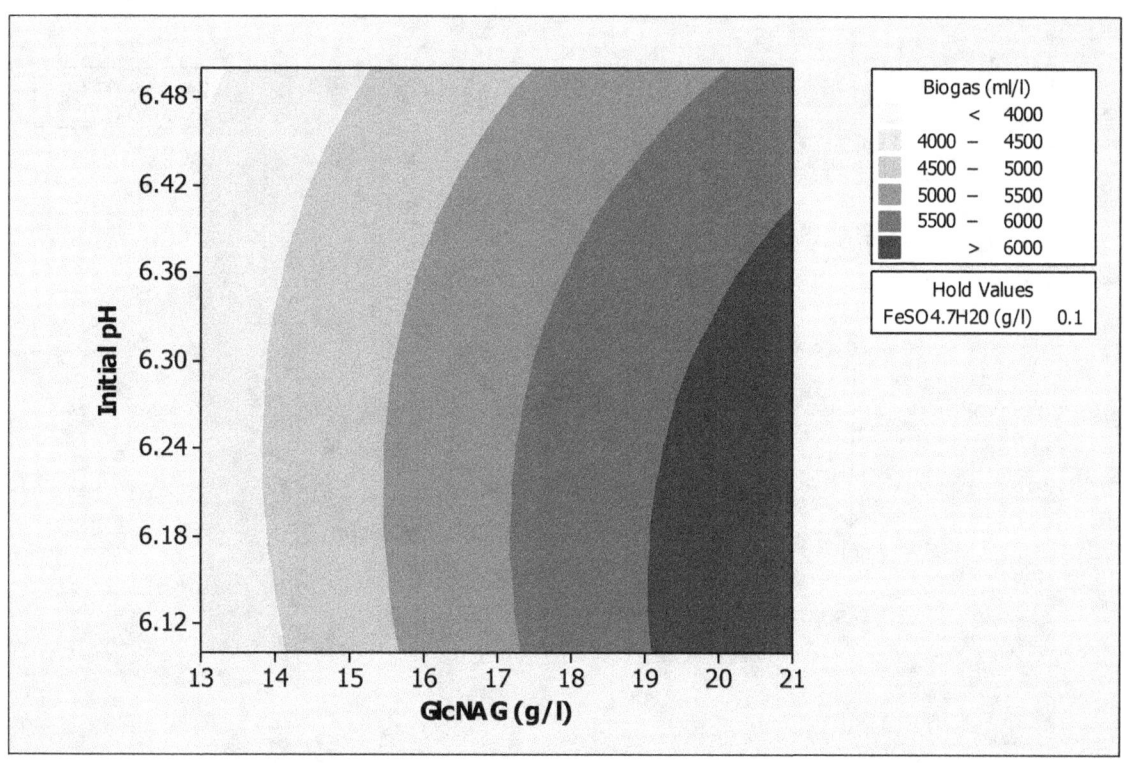

Figure 6.4. *Clostridium beijerinckii* biogas production response surface plots showing the effect of initial N-acetylglucosamine concentrations and initial pH with optimum level of $FeSO_4.7H_2O$ of 0.1 g/l.

Figure 6.3 and 6.4 show the response surface plot and resultant contour plots for independent variables N-acetylglucosamine (X1) and initial pH (X3) whilst the third variable $FeSO_4.7H_2O$ (X2) was constant at its optimal level. There was a substantial increase in total biogas volume when the concentration of N-acetylglucosamine increased. Total biogas also increases as initial pH was reduced and this effect was more pronounced within the ranges of 19 and 21 g/l of N-acetylglucosamine. The result suggests that total volume of biogas produced can be maximized at a lower initial pH within the N-acetylglucosamine ranges of 19 to 21 g/l. However there was a small interaction effect between these variables but the effect was insignificant (P > 0.05).

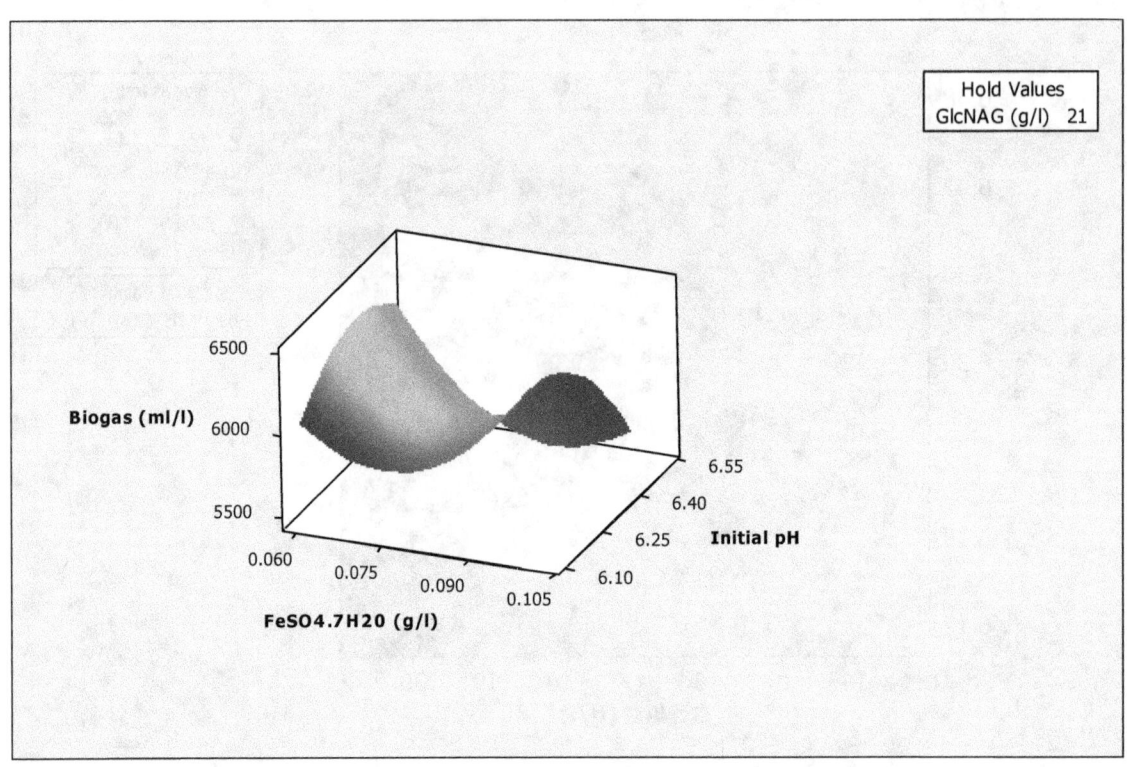

Figure 6.5. *Clostridium beijerinckii* biogas production response surface plot showing the effect of $FeSO_4.7H_2O$ and initial pH with optimum level of N-acetylglucosamine (21 g/l).

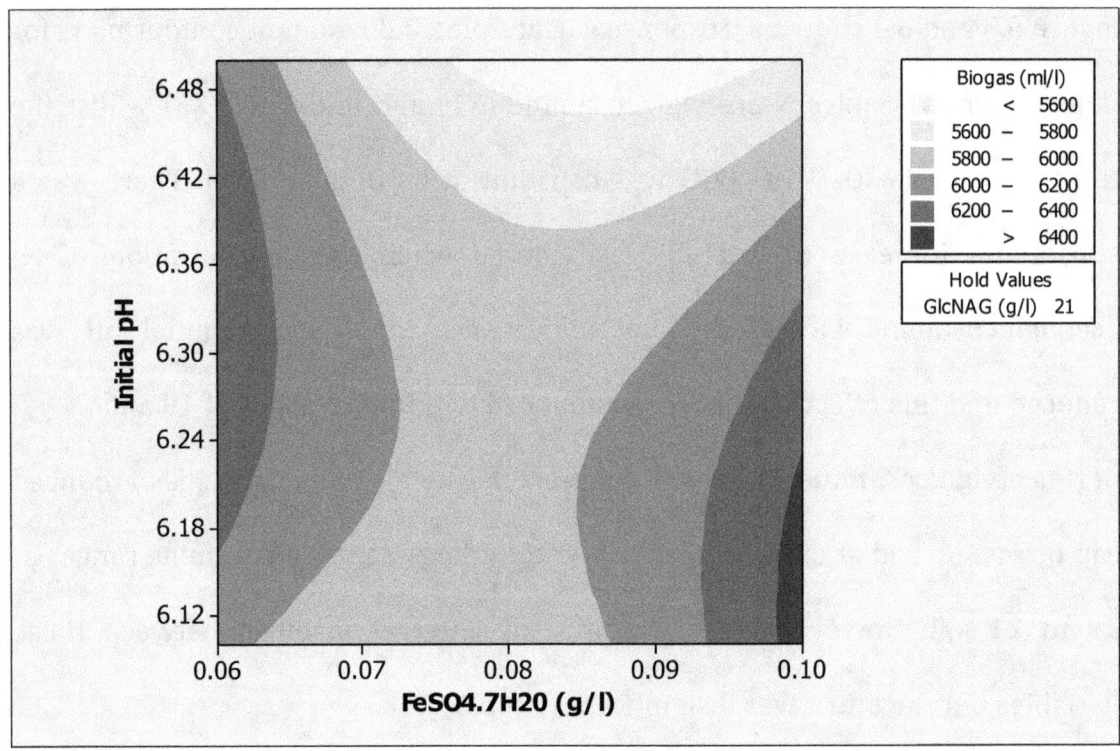

Figure 6.6. *Clostridium beijerinckii* biogas production contour plot showing the effect of $FeSO_4.7H_2O$ and initial pH with optimum level of N-acetylglucosamine (21 g/l).

Figure 6.5 and 6.6 show the response surface plot and resultant contour plots for independent variables FeSO$_4$.7H$_2$O (X2) and initial pH (X3) while the third variable N-acetylglucosamine (X1) was constant at its optimal level. These plots represent a minimax response surface. From the saddle point near the center of the contour plot, increasing or decreasing the two variables simultaneously leads to a decrease in total biogas produced. However increasing any of the variables while decreasing the other leads to an increase in total biogas produced. To maximize the total biogas produced, settings for FeSO$_4$.7H$_2$O are around 0.10 g/l and for initial pH around 6.12. There were interaction effects between these two variable but these effects were not significant ($P > 0.124$).

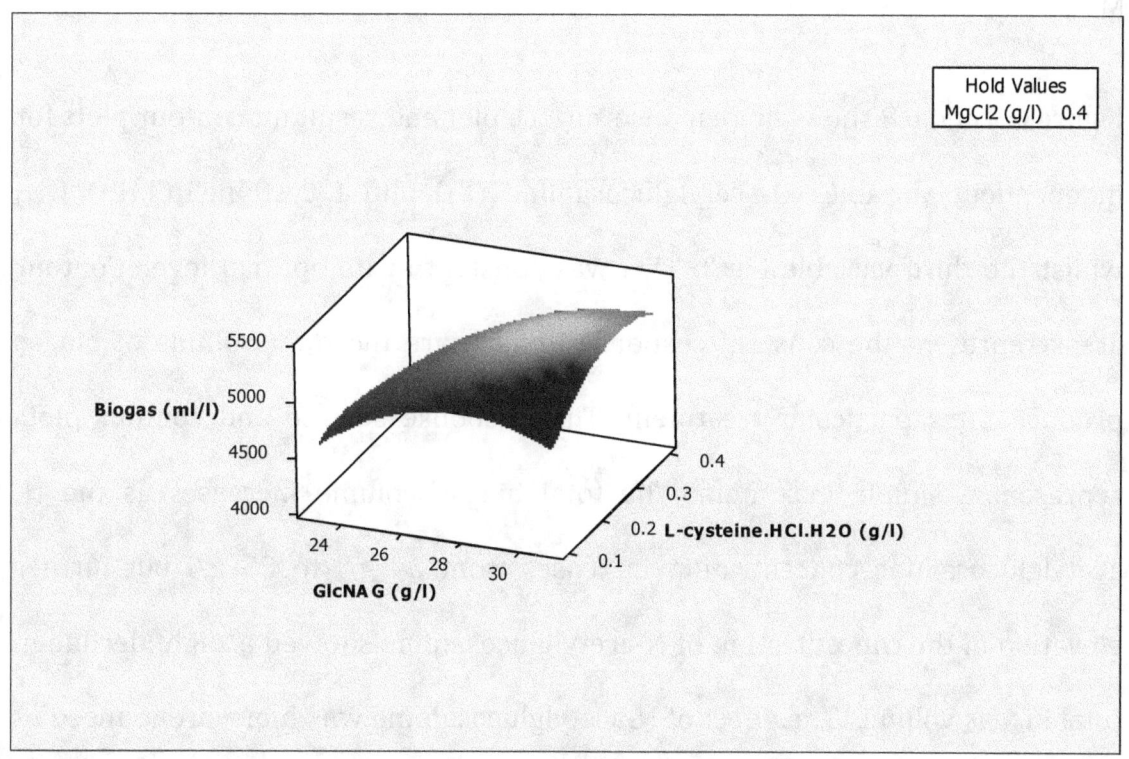

Figure 6.7. *Clostridium paraputrificum* biogas production response surface plot showing the effect of initial N-acetylglucosamine concentrations and initial L-cysteine.HCl.H$_2$O concentrations with optimum level of MgCl$_2$

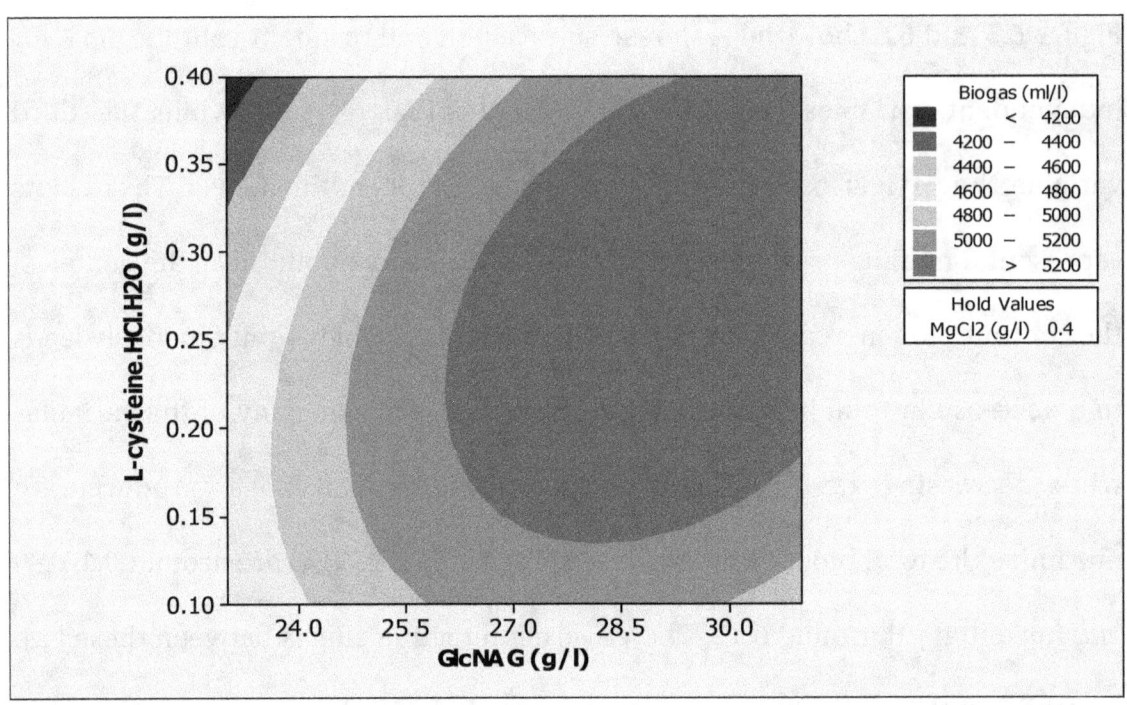

Figure 6.8 *Clostridium paraputrificum* biogas production contour plots showing the effect of N-acetylglucosamine and L-cysteine.HCl.H_2O with optimum level of $MgCl_2$

Figure 6.7 and 6.8 show the response surface plot and resultant contour plots for independent variables N-acetylglucosamine (X_1) and L-cysteine.HCl.H_2O (X_2) whilst the third variable $MgCl_2$ (X_3) was constant at its optimal level. Contour areas represent the constant responses, which are the total volume of biogas produces at a particular treatment. This response surface and contour plots represent a simple maximum. The total biogas volume increases as the N-acetylglucosamine concentration increases from 24 g/l to 29 g/l but further elevation of the concentration of N-acetylglucosamine showed a slight decline in total biogas volume. The effect of N-acetylglucosamine was more pronounced as L-cysteine.HCl.H_2O concentration increased. The two variables were slightly interdependent and there were interaction effects between the two variables but were insignificant ($P > 0.069$).

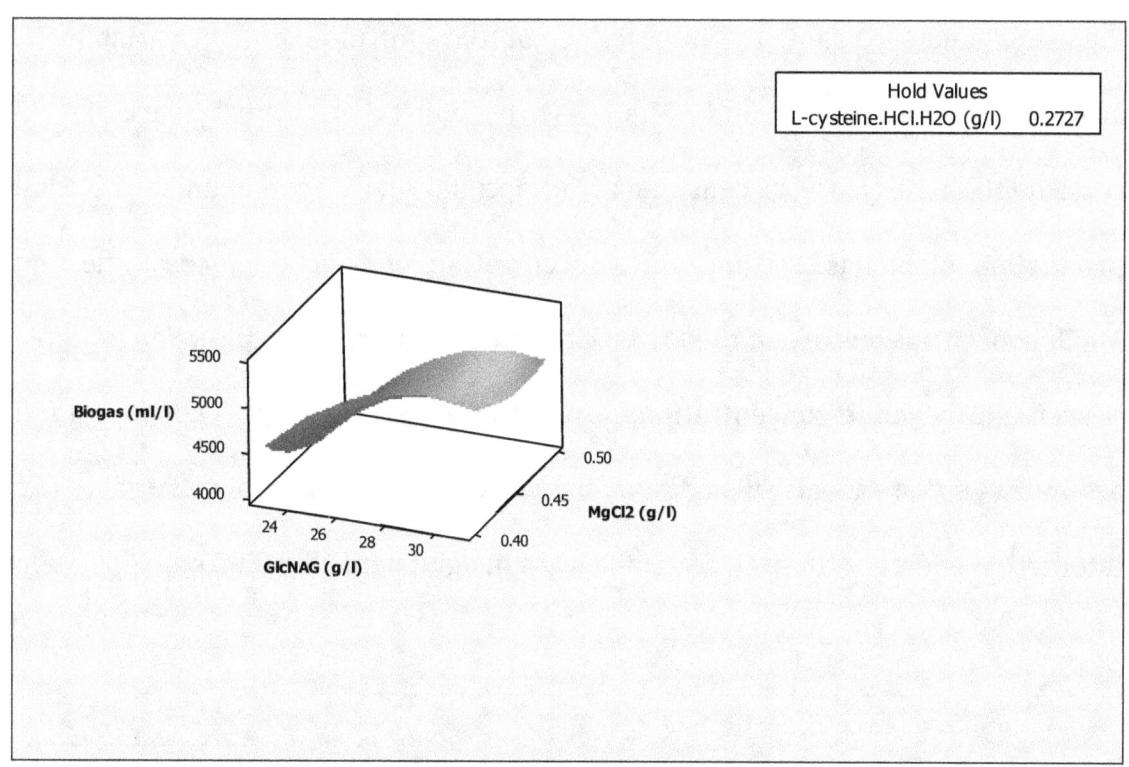

Figure 6.9. *Clostridium paraputrificum* biogas production response surface plot showing the effect of initial N-acetylglucosamine concentrations and initial $MgCl_2$ concentrations with optimum level of L-cysteine.HCl.H_2O.

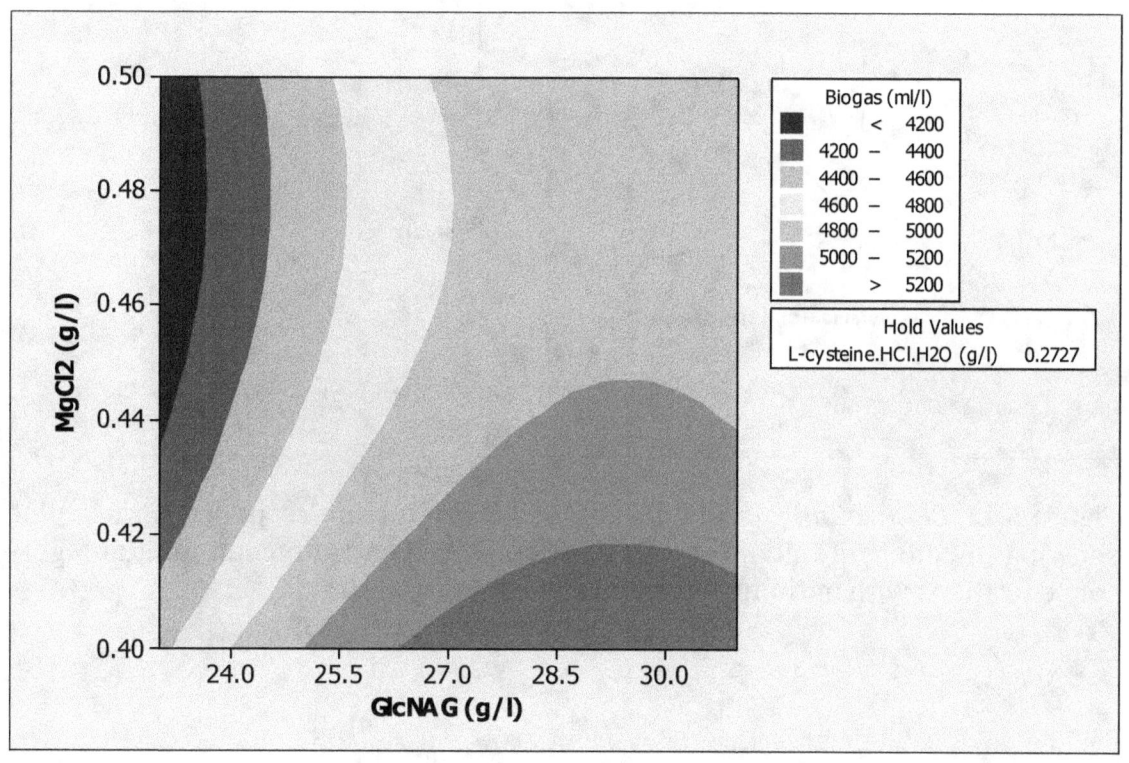

Figure 6.10. *Clostridium paraputrificum* biogas production contour plots showing the effect of initial N-acetylglucosamine concentrations and initial $MgCl_2$ concentrations with optimum level of L-cysteine.HCl.H_2O.

Figure 6.9 and 6.10 show the interacting contours for independent variables N-acetylglucosamine (X_1) and $MgCl_2$ (X_3) whilst the third variable L-cysteine.HCl.H_2O (X_2) was constant at its optimal level. To maximize the total production of biogas, settings for concentration of N-acetylglucosamine and $MgCl_2$ concentrations are in the lower right side of the contour plot. For example N-acetylglucosamine concentrations around 29.25 g/l and $MgCl_2$ concentration around 0.4 g/l are ideal. This contour matches a minimax response surface. The interaction effects of these two variables are insignificant ($P > 0.05$).

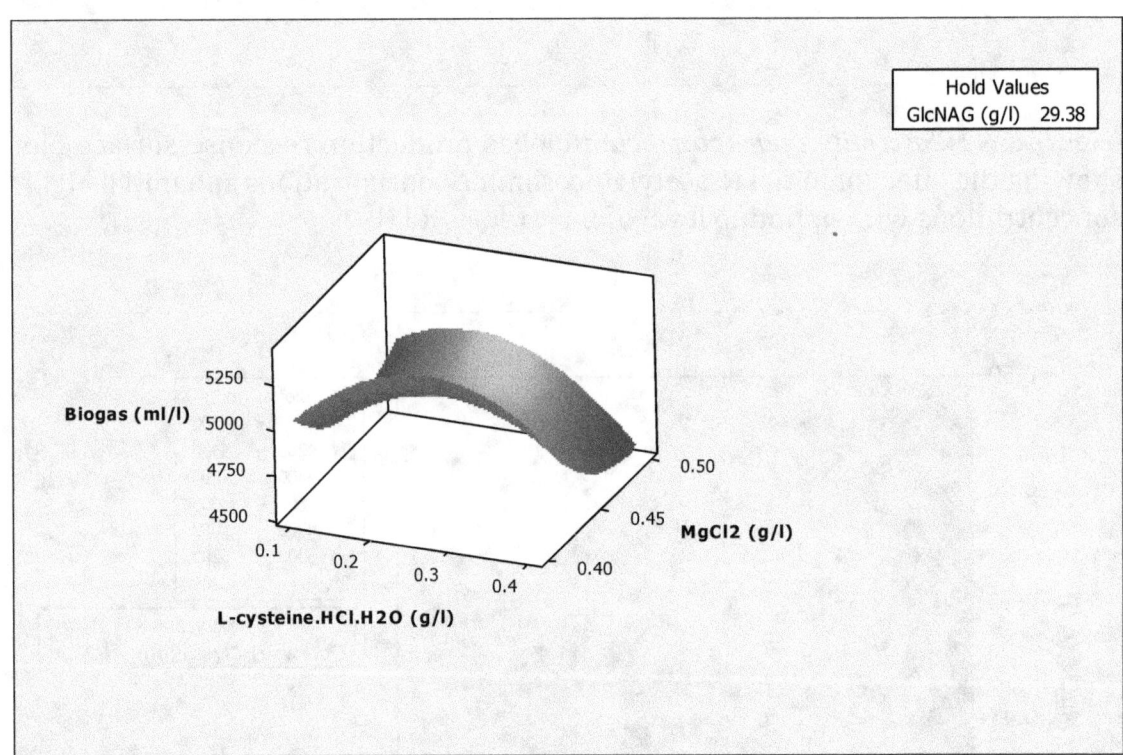

Figure 6.11. *Clostridium paraputrificum* biogas production response surface plot showing the effect of initial L-cysteine.HCl.H_2O concentrations and initial $MgCl_2$ concentrations with optimum level of N-acetylglucosamine.

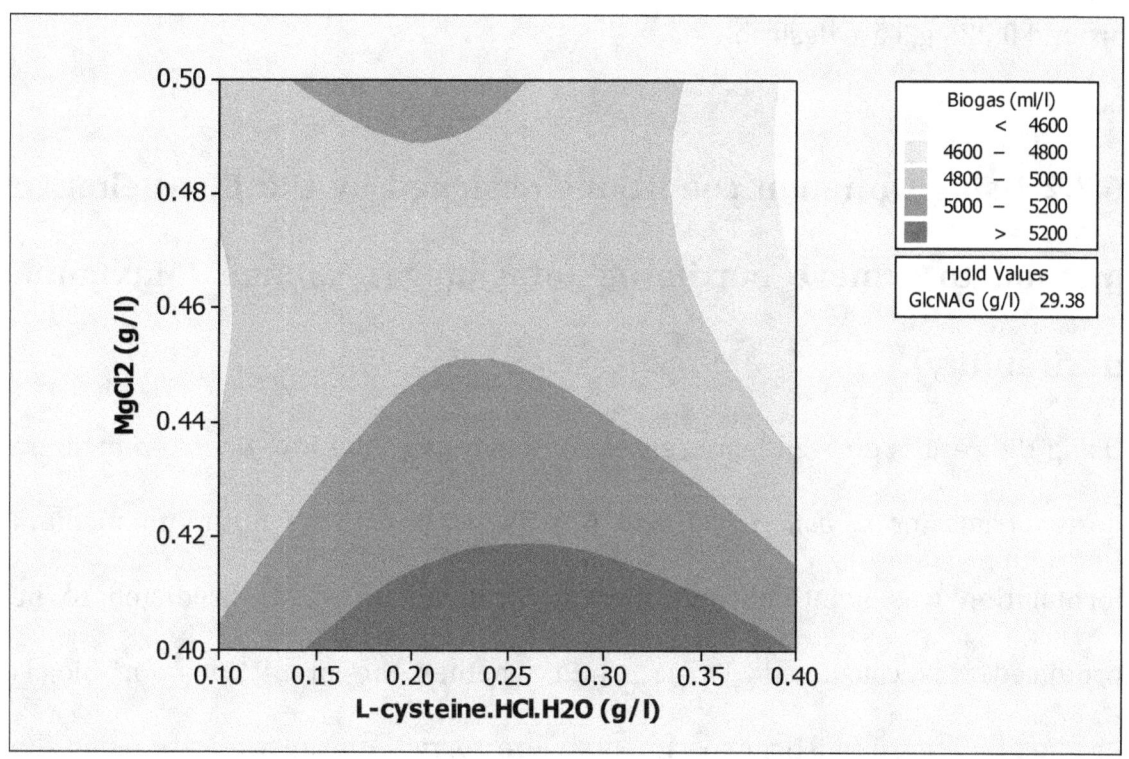

Figure 6.12. *Clostridium paraputrificum* biogas production contour plots showing the effect of initial L-cysteine.HCl.H$_2$O concentrations and initial MgCl$_2$ concentrations with optimum level of N-acetylglucosamine.

Figure 6.11 and 6.12 show the response surface plot and resulting contour plots for independent variables L-cysteine.HCl.H$_2$O (X$_2$) and MgCl$_2$ (X$_3$) while the third variable N-acetylglucosamine (X$_1$) was constant at its optimal level. The response surface and contour plots represent a minimax response surface. From the saddle point close to the center of the design, increasing or decreasing L-cysteine.HCl.H$_2$O and MgCl$_2$ simultaneously results in a decrease in the volume of total biogas produced. Although from the saddle point also known as the stationary point, increasing either of the independent variable while decreasing the other variable results in an increase in the total biogas produced. The interaction effect of L-cysteine.HCl.H$_2$O and MgCl$_2$ concentrations were insignificant (P > 0.05). However to achieve the highest total volume of biogas, MgCl$_2$ concentrations around 0.4 g/l and for L-cysteine.HCl.H$_2$O concentration around 0.275 g/l are desirable.

6.3.2 Using Optimum conditions designed by the Box-Behnken method to achieve maximum total biogas volume (Maximum desirability).

Using the results derived from regression analysis of the Box-Behnken medium design trials for *C. beijerinckii* and *C. paraputrificum*, the optimum medium formulation was identified and the total volume of biogas predicted to be produced was calculated. To ascertain whether the predictions for biogas production were realistic, experiments were carried out in triplicate using the optimised medium formulation for each species. Using the optimized medium formulation conditions, the maximum total volumes of biogas predicted were 6478.95 ml/l for *C. beijerinckii* and 5397.78 ml/l for *C. paraputrificum* the actual amounts of biogas produced were 6533 ml/l for *C. beijerinckii* and 5350 ml/l for *C. paraputrificum*. The metabolites produced by the *C. beijerinckii* cultures were acetate and butyrate whilst those produced by the *C. paraputrificum* cultures were acetate, butyrate and lactate (Table 6.11). Table 6.11 shows the concentration of metabolites produced, N-acetylglucosamine utilized as well as total volume of biogas produced for *C. beijerinckii* and *C. paraputrificum* cultures.

Table 6.11. Final biomass, N-acetylglucosamine consumption, biogas production volume and yield of products using optimum conditions predicted by the Box-Behnken design for *C. beijerinckii* and *C. paraputrificum.*

Clostridium Species	Final biomass (g/l)	GlcNAG utilised (g/l)	Final pH	Yield of products (g/l)				Biogas Volume (ml/l)
				Lactate	Acetate	Ethanol	Butyrate	
Clostridium beijerinckii	2.47	18.08	5.23	0.00	4.78	0.00	4.43	6533
Clostridium paraputrificum	1.72	19.59	5.09	2.58	6.75	0.45	3.74	5350

g/l = grams per litre

ml/l = millilitres of biogas produced per litre of medium used

6.3.3 Fermentation profile of *C. beijerinckii* and *C. paraputrificum* using the optimised medium formulation in batch fermentations using a 1 litre bioreactor.

The characteristics of the biogas fermentation by *C. beijerinckii* and *C. paraputrifiucum* under optimized conditions were carried out, based on the results of the statistically designed experiments; the optimized medium was prepared as shown in table 6.4. Table 6.12 and 6.13 demonstrate the time course profiles for *C. beijerinckii* and *C. paraputrificum* fermentation run with N-acetylglucosamine as the main carbon source under optimized conditions. The organic acids detected from the fermentation run were butyrate and acetate for *C. beijerinckii* cultures and lactate, actetate and butyrate for *C. paraputrificum* cultures. Butyrate was the most abundant organic acid produced in the *C. beijerinckii* ($P < 0.036$) and also most abundant at all time points during the course of fermentation ($P < 0.005$) while acetate was the most abundant organic acid ($P < 0.01$) in the *C. paraputrificum* cultures followed by butyrate and then lactate. At all time points acetate was significantly the most abundant organic acid produced (P < 0.01). These results were obtained at the end of the fermentation run when the utilization of N-acetylglucosamine ended.

The pH of the culture medium declined over the course of the fermentation, for each of the three replicates, from an initial mean pH of 5.99 to a final one of 5.37 for *C. beijerinckii* and from an initial mean pH of 6.20 to a final one of 4.98 for *C. paraputrificum*. There was also a significant reduction in the residual N-acetylglucosmine concentration in the culture medium. The fermentation lasted for duration of 75 hours for *C. beijerinckii* and 58 h for *C. paraputrificum* cultures. Table 6.12 and 6.13 shows the profile of the fermentation run for *C. beijerinckii* and *C. paraputrificum* respectively. It shows the concentration of organic acids produced and residual N-acetylglucosamine concentrations.

Table 6.12 Fermentation profile for *C. beijerinckii* relating the pH of the medium and the residual N-acetylglucosamine with biomass, and organic acid yield over the 75 h course of the fermentation.

Cultivation Time (h)	Culture pH	Residual GlcNAc (g/l)	Biomass concentration (g/l)	Yield of products (g/l)	
				Acetate	Butyrate
0	5.99	20.15	0.19	0.00	0.00
4	5.97	19.86	0.19	0.00	0.00
21	5.47	10.54	2.04	2.64	3.78
25	5.45	8.01	2.52	3.03	4.50
28	5.42	6.38	2.77	3.22	4.64
31	5.41	4.89	2.90	3.49	4.93
49	5.40	0.00	2.95	4.26	6.05
73	5.39	0.00	2.72	4.33	6.00
75	5.37	0.00	2.74	4.37	6.01

g/l = grams per liter
h = hour
n = 3

Table 6.13 Fermentation profile for *C. paraputrificum* relating the pH of the medium and the residual N-acetylglucosamine with biomass, and organic acid yield over the 58 h course of the fermentation.

Cultivation Time (h)	Culture pH	Residual GlcNAc (g/l)	Biomass concentration (g/l)	Yield of products (g/l)		
				Lactate	Acetate	Butyrate
0	6.20	28.68	0.18	0.00	0.00	0.00
2	6.18	28.32	0.20	0.00	0.00	0.00
6	6.16	28.20	0.21	0.00	0.00	0.00
8	6.16	28.04	0.23	0.00	0.00	0.00
26	5.16	15.26	1.36	2.31	4.73	2.63
30	5.10	13.99	1.43	2.59	5.15	2.83
35	5.07	13.37	1.31	2.84	5.47	2.85
38	4.98	13.28	1.14	2.83	5.38	2.85
48	4.99	13.32	1.19	2.84	5.39	2.83
58	4.98	13.21	1.19	2.80	5.36	2.84

g/l = grams per liter
h = hour
n = 3

Table 6.14 Yield coefficients of organic acid products, carbon recovery and theoretical hydrogen yield from batch cultures of *C. beijerinckii* over the 75 h course of the fermentation.

Cultivation Time (h)	Yield coefficient of product (mol of product/mole GlcNAc consumed)		Carbon recovery from GlcNAc (%)	Theoretical yeild of Hydrogen gas. (mol of H_2/mol GlcNAc)
	Acetate	Butyrate		
0	0.00	0.00	0.00	0.00
4	0.00	0.00	0.00	0.00
21	1.03	1.00	75.76	2.45
25	0.94	0.94	70.38	2.81
28	0.88	0.86	64.72	3.19
31	0.86	0.82	62.45	3.34
49	0.79	0.76	57.90	3.64
73	0.81	0.76	57.96	3.64
75	0.81	0.76	58.19	3.63

g/l = grams per liter
h = hour
% = Percentage
n = 3

Table 6.15 Yield coefficients of organic acid products, carbon recovery and theoretical hydrogen yield from batch cultures of *C. paraputrificum* over the 58 h course of the fermentation.

Cultivation Time (h)	Yield coefficient of product (mol of product/mole GlcNAc consumed)			Carbon recovery from GlcNAc (%)	Theoretical yeild of Hydrogen gas. (mol of H_2/mol GlcNAc)
	Lacate	Acetate	Butyrate		
0	0.00	0.00	0.00	0.00	0.00
2	0.00	0.00	0.00	0.00	0.00
6	0.00	0.00	0.00	0.00	0.00
8	0.00	0.00	0.00	0.00	0.00
26	0.43	1.32	0.50	50.56	2.71
30	0.44	1.31	0.49	50.86	2.73
35	0.46	1.34	0.47	52.34	2.69
38	0.46	1.31	0.47	51.47	2.75
48	0.46	1.31	0.47	51.65	2.74
58	0.45	1.30	0.47	50.90	2.80

g/l = grams per liter.
h = hour.
% = Percentage.
n = 3

Table 6.16 Yield coefficient of moles of biogas per mole of N-acetylglucosamine consumed, percent of H_2 and CO_2 gas expelled from 1 litre batch cultures of *C. beijerinckii* and *C. paraputrificum* at optimum conditions using the ideal gas law assuming conditions of temperature at 37°C and pressure at 1 atm.

Clostridium Species	Yield coefficient of product (biogas) per GlcNAc consumed (mol/mol)			H_2 gas expelled (%)	CO_2 gas expelled (%)
	Biogas	H_2	CO_2		
Clostridium Beijerinckii	2.85	2.43	0.42	87.08	12.91
Clostridium Paraputrificum	3.04	2.55	0.49	85.76	14.24

% = Percentage

There was an increase in biomass concentration as the fermentation run proceeded with a final biomass concentration of 2.74 g/l (*C. beijerinckii*) and 1.19 g/l (*C. paraputrificum*). The total volume of biogas produced was 6610 ml/l (*C. beijerinckii*) and 5405 ml/l (*C. paraputrificum*). Table 6.14 and 6.15 show the yield coefficient of the metabolized products per utilized N-acetylglucosamine as

212

well as the recovery of carbon throughout the cultivation period for *C. beijerinckii* and *C. paraputrificum* respectively.

In the batch fermentation run the N-acetylglucosamine concentration in the medium, cell biomass concentration and pH changed according to time duration. The carbon recovery was however calculated in percentages as the total amount in moles of carbon contained in the organic acids produced per mole of N-acetylglucosamine consumed. As fermentation proceeded there was a decline in residual N-acetylglucosamine concentration and the maximum theoretical yield of hydrogen was 3.64 mol H_2/ mol of N-acetylglucosamine for *C. beijerinckii* and for *C. paraputrificum* 2.80 mol H_2/ mol of N-acetylglucosamine and these theoretical values were based on substrate consumed and organic acids produced. However assuming conditions are at 37°C and pressure at 1 atm using the ideal gas law, a maximum of 2.43 mol of H_2 / mol of N-acetylglucosamine for *C. beijerinckii* and 2.55 mol of H_2 / mol of N-acetylglucosamine for *C. parautricicum* cultures were achieved (Table 6.16).

6.4 Discussion

The Box-Behnken method was used to optimize the medium formulation to achieve maximum total biogas produced from N-acetylglucosamine by both *C. beijerinckii* and *C. paraputrificum*. Both *C. beijerinckii* and *C. paraputrificum* metabolised the N-acetylglucosamine and produced organic acids, which led to a significant reduction in the pH of the medium (Tables 6.5 and 6.6). Acetate and butyrate were the major organic acid products of *C. beijerinckii* whilst lactate, acetate, formate and butyrate were the major organic acid products of *C. paraputrificum*. In addition, some ethanol was also produced by the *C. paraputrificum* cultures. Interestingly, changes in the pH of the culture medium may influence the species of organic acids produced as fermentation proceeds as it can influence the metabolic pathways used by the bacterium. For example Zhu and Yang (2004) reported that the pH of the culture medium for *Clostridium tyrobutyricum* not only influenced bacterial cell growth and the fermentation rate, but it also led to changes in the yield and the range of metabolic products evolved. When the culture medium pH was 6.0 butyrate was the most abundant organic acid produced by *C. tyrobutyricum*, but when the culture medium pH was changed to a pH 5.0 lactate and acetate were the most abundant organic acids.

In the experiments described in this work when visible production of biogas by the *C. paraputrificum* cultures ceased, there was still residual N-acetylglucosamine in all of the experiments and the final pH ranged from 5.20 to 4.99. It is possible that the accumulation of organic acids, and the resulting decrease in the pH of the growth medium inhibited N-acetylglucosamine utilization and biogas production when the concentration of organic acid exceeded a critical limit. The effect caused by pH of a culture medium is due to the change in ionization state of components involved in enzymatic reactions (Fabiano & Perego, 2002). Van Ginkel and Logan (2005) concluded that low initial culture medium pH and high concentrations of organic acids during fermentation inhibit bio-hydrogen production. Low initial culture medium pH values of 4.0 - 4.5 cause longer lag periods (Khanal et al., 2004). Zhang et al., (2003) reported that high culture medium initial pH of 9, decrease lag time, but

214

however produce a lower yield of hydrogen.

For *C. beijerinckii* the greatest volume of biogas was produced from cultures, which used the greatest amount of N-acetylglucosamine. However it would seem that the quantity of biogas produced is not solely dependent upon the quantity of N-acetylglucosamine used as the *C. paraputrificum* cultures that yielded the greatest total volume of biogas had not used the greatest amount of N-acetylglucosamine amongst all the cultures. The total volume of biogas produced also depends on the influence of other variables in the medium design. Trial design T_5 and T_7 (Table 6.5) in the *C. beijerinckii* cultures and T_3 (Table 6.6) in the *C. paraputrificum* cultures had the lowest total volume of biogas produced 3867 ml/l, 3750 ml/l and 3433 ml/l respectively, the low quantity of N-acetylglucosamine utilized (12.38 g/l, 12.87 g/l and 18.16 g/l) resulted in the low volume of biogas produced. These results show that these medium formulations relating to low total volume of biogas produced were not favorable to optimal biogas production. The quantity of carbon source consumed by the bacteria during a fermentation reaction may also influence the yield of products.

N-acetylglucosamine was used successfully as a carbon source by both *C. beijerinckii* and *C. paraputrificum* as shown by the production of biomass, organic acids and biogas. When a bacterium metabolizes a suitable carbon source, high-energy phosphate compounds such as ATP are formed which then permit the bacterium to undertake its normal metabolic activities (Greasham & Herber, 1997). Therefore if the cells were unable to use N-acetylglucosamine as a carbon source in the current experiments, it would lead to shortages of energy in the form of ATP and this would result in little or no biogas production, as the cells would effectively be starved. Consequently, when seeking to design a medium formulation that aims to maximize the production of a target metabolite it is necessary to determine the optimum concentration of the chosen carbon source to include. It is important to use a suitable concentration of carbon source for the target product. Thus the optimum concentration of N-acetylglucosamine to maximize the biogas production by cultures of *C. beijerinckii* and *C. paraputrificum* was identified by using the Box-Behnken method. The optimal level of the key medium constituent factors for *C. beijerinckii* cultures (N-acetylglucosamine, $FeSO_4.7H_2O$ and initial pH) and *C. paraputrificum* cultures (N-acetylglucosamine, L-cysteine.$HCl.H_2O$ and $MgCl_2$) as well as the effect of their interactions on total biogas produced were further explored.

The regression model F-values were 13.97 (*C. beijerinckii*) and 9.18 (*C. paraputrificum*). These results were significant. The statistical result; test the overall significance of the regression model. The p-values for *C. beijerinckii* 0.005 and *C. paraputrificum* 0.013 for the corresponding F-values were less than 0.05. The significance of the regression model also test whether the terms (medium constituents) in the model have any effect on the response (total biogas produced) which is the total biogas produced. That is, at least one of the terms in the regression equation, (equation 6.1 for *C. beijerinckii* and 6.2 for *C. paraputrificum*) has an impact on the mean response. The P-values are used as a tool to determine which of the effects in the regression model are statistically significant, which also shows the strength of interaction between each individual variable (Liu et al., 2003). The smaller the resulting P-values, the greater the significance of the corresponding independent variable (Muralidhar et al., 2001). P-values less than 0.05 indicate that the regression model terms are significant. In this case, coded values X_1 and X_2^2 (*C. biejerinckii*) and X_1, X_2, X_3, X_1^2, X_2^2 (*C. paraputrificum*) are significant model terms as shown in tables 6.9 and 6.10 respectively. P-values greater than 0.05 indicate the model terms are not significant.

R^2 values were 0.9618 (*C. beijerinckii*), and 0.9429 (*C. paraputrificum*); these values represent the proportion of variation in the total volume of biogas produced that is described in the model. In other words the R^2 values explain 96.18 % for *C. beijerinckii* and 94.29 % for *C. paraputrificum* variability of the response (Total volume of biogas produced) that is attributed to the independent variables involved. It also indicates a significant agreement between experimental and predicted values. This implies that the mathematical model is very reliable for total biogas production in this analysis.

Analyzing interaction effects for *C. beijerinckii* cultures the terms X_1X_2, X_1X_3 and X_2X_3 were considered insignificant. That is the effect of N-acetylglucosamine concentrations on total biogas volume produced was independent of $FeSO_4.7H_2O$ concentrations and initial pH ranges. The effect of $FeSO_4.7H_2O$ concentrations on total biogas volume was also independent of initial pH ranges used in this study. Jo et al., (2008) reported that there were insignificant interaction effects between glucose concentration and pH for bio-hydrogen production by *C. tyrobutyricum*. Furthermore using the response surface methodology, the interaction effect of initial pH, glucose and $FeSO_4$ concentrations on bio-hydrogen production yield by *C. butyricum* EB6 were analysed, Chong et al., (2009) concluded that the interaction effect between glucose concentration and $FeSO_4$ concentration were insignificant and also the interaction effect between glucose and pH were insignificant. For *C. paraputrificum* cultures interaction effects terms X_1X_2, X_1X_3 and X_2X_3 were also considered insignificant, meaning the effects of N-acetylglucosamine concentrations on the total volume of biogas produced were independent of the L-cysteine.HCl.H_2O and $MgCl_2$ medium concentrations. The effects of L-cysteine.HCl.H_2O concentrations on the total volume of biogas produced were also independent of the $MgCl_2$ medium concentrations. Interaction effects of glucose and magnesium were insignificant in experiments conducted by Guo et al., (2009) for bio-hydrogen yield.

The resulting square terms effects are used to determine whether or not there are curvature or quadratic effects in the response surface. *Clostridium beijerinckii* squared term for X_2^2 was significant as shown in table 6.9 ($P < 0.035$) and was treated individually. Squared effects for X_1^2 ($P > 0.413$) and X_3^2 ($P > 0.165$) showed insignificant quadratic effects. That is the relationship between N-acetylglucosamine and total biogas produced, and initial pH and total biogas produced follow a straight line on the other hand the relationship between $FeSO_4.7H_2O$ and the total biogas produced follow a curved line. Iron concentration in a culture medium is also an important medium constituent to be included in a medium design as too high or too low iron concentration may have a significant effect on biogas produced. Bio hydrogen production is facilitated by intracellular hydrogenase enzyme. The iron containing Fe-hydrogenase enzyme catalyzes the oxidation of molecular hydrogen from protons and electrons. This process of catalyzing the production of hydrogen is a reversible reaction (Nicolet, Cavazza & Fontecilla-Camps, 2002). Explaining the squared term effect of $FeSO_4.7H_2O$ on total biogas produced within the experimental ranges, a point was achieved were further increases in $FeSO_4.7H_2O$ above an optimum level for biogas production resulted in a decline in biogas produced. Wang and Wan (2008), investigated the effect of different Fe^{2+} concentrations on biohydrogen production by mixed cultures. Biohydrogen production increased as Fe^{2+} concentration increased from 0 to 0.25 g/l but decreased as Fe^{2+} concentrations were above 0.25 g/l. In this study 0.02 g/l of Fe^{2+} (0.1 g/l $FeSO_4.7H_2O$) was the optimum concentration needed for production of the highest total volume of biogas for *C. beijerinckii* cultures.

For squared term analysis for *C. paraputrificum* cultures X_1^2 and X_2^2 with p-values of 0.019 and 0.038 had significant evidence of a quadratic effect. Pan et al., (2008) also recorded a significant squared effect of glucose during a bio-hydrogen optimization by *Clostridium sp.* Fanp2. In this study this indicates that the relationship between N-acetylglucosamine and total biogas produced, and L-cysteine.HCl.H₂O and total biogas produced follow a curved line effect. The X_3^2 squared effect is insignificant (P = 0.095). This indicates that the relationship between $MgCl_2$ and total biogas produced followed a straight line. The linear line in the relationship between the variable and the total biogas produced indicates that as the value of the independent variable increases the resulting effect increases. However in the curved line increasing an independent variable infinitely, a point will be reached whereby the response does not increase infinitely. At some point a quadratic relationship can occur, with an intermediate optimum, which can signify a maximum or a minimum subject to the desired response.

This model contains 3 linear effects (X_1, X_2, X_3) each for *C. beijerinckii* and *C. paraputrificum*. The p-value of 0.0 for X_1 in the *C. beijerinckii* cultures shows a significant linear effect for N-acetylglucosamine. Meaning the total biogas produced differs depending on the concentrations of N-acetylglucosamine. Pan et al., (2008) reported significant linear effects of glucose on bio-hydrogen production, a similar result was achieved by Jo et al., (2008), as glucose and culture pH were detected to have significant linear effects on bio-hydrogen production by *C. tyrobutyricum* cultures. However in this study linear effect for X_2 ($FeSO_4.7H_2O$) and X_3 (Initial pH) respectively were not significant on total biogas produced. For linear effects of *C. paraputrificum* cultures X_1 (N-acetylglucosamine), X_2 (L-cysteine.$HCl.H_2O$) and X_3 ($MgCl_2$) have significant effects. This means the total volume of biogas produced differs depending on the individual concentrations of N-acetylglucosamine, L-cysteine.$HCl.H_2O$ and $MgCl_2$. $MgCl_2$ was also identified to be significant amongst other medium components investigated; its optimum concentration was 0.4 g/l in the *C. paraputrificum* cultures. Magnesium ions are essential and involved in many enzyme activities; they also function to stabilize cell membrane. Zhao et al. (2012) stated that Mg^{2+} influenced bacteria cell growth and production of hydrogen by *C. beijerinckii* RZF-1108. They suggested that *C. beijerinckii* RZF-1108 cell growth was dependent on the presence of Mg^{2+} ions in the culture medium. Bio hydrogen production also increased with the increase in Mg^{2+} concentration. Inhibition of hydrogen production was noted when Mg^{2+} concentration was over 0.4 g/l. They also studied the effect of L-cysteine on hydrogen production and concluded that L-cysteine is not an ideal stimulus for production of bio-hydrogen by *C. beijerinckii* RZF-1108, as the maximum bio hydrogen yield and biomass were achieved when L-cysteine was absent. Although L-cysteine is a reducing agent supplied in anaerobic culture media to reduce the oxidation-reduction potential, in mixed cultures. Thus identified as a significant constituent in the Box-Behnken design.

A large number of experiments are usually involved when choosing a fermentation improvement strategy. This proliferation of experiments necessitates the use of small-scale experiments, like the shake flask cultures or syringe cultures used in the experiments described in this work, because space and resources are usually limited. However there are some drawbacks to small-scale systems for example it is not possible to include proper control systems for pH, gas transfer between medium and atmosphere can be poor, evaporation from the medium can be a significant proportion of the total volume and mixing of medium can be inadequate leading to uneven distribution of nutrients (Kennedy et al., 1994). Nevertheless researchers normally assume that an optimized medium chosen on the basis of shake flask culture results will remain the best medium in a large-scale stirred tank too (Kennedy & Krousereff, 1999). Consequently, the optimized medium formulation derived from the statistical analysis of the results of the Box-Behnken method results was tested in a bioreactor scaled up to a working volume of 1 litre.

The anaerobic production of biogas from a carbon source also leads to the co-production of organic acids and alcohols (Hawkes et al., 2002). In table 6.12 and 6.13 at time intervals, the culture medium pH decreased, corresponding with the increase in product formation of acidic metabolites. This implies that pH had an important effect on the concentration of organic acids produced. In this study, the decrease of the culture medium final pH irrespective of the initial pH, correlated with increases in organic acid production. The metabolic products formed are as a result of changes in the metabolic pathway of the microorganisms (Evvyernie et al., 2000). In the current experiments, the bacterial cultures were able to consume N-acetylglucosamine from the growth medium to yield biomass, biogas and organic acids. N-acetylglucosamine was completely consumed by the *C. beijerinckii* cultures after 31 h of cultivation. It appears that exhaustion of the N-acetylglucosamine supply led to the cessation of fermentation by *C. beijerinckii* as there were no further significant changes in the production of organic acids, the pH of the culture medium, or the biomass concentration once the N-acetylglucosamine supply was spent. In contrast, the low final pH of the culture medium seems to have been responsible for the cessation of fermentation by the *C. paraputrificum* cultures as there was residual N-acetylglucosamine left in the culture medium after 35 hours even though there were no further significant changes in biomass, residual N-acetylglucosamine concentration in the medium or organic acid concentration. A low initial pH in the growth medium, and a low pH during fermentation usually lead to a low level of intracellular ATP resulting in inhibition of bacterial growth (Bowles & Ellefson, 1985). Since inhibition of bacterial cell growth will eventually lead to an end to the production of metabolites or target responses, this may be a possible explanation for the cessation of fermentation by *C. paraputrificum* before all of the N-acetylglucosamine had been used. Indeed, it has been observed previously that a low initial culture medium pH and the accumulation of high concentrations of organic acids during fermentation both inhibit bio-hydrogen production (Khanal et al., 2004; Van Ginkel and Logan, 2005; Zhao et al., 2011).

Optimization of the medium formulation led to an increase in the total volume of biogas produced by *C. beijerinckii*, from 5733 ml/l during the Method of Steepest Ascent experiments to 6533 ml/l (from the syringe cultures) and 6610 ml/l (from bioreactor batch cultures) during the Box-Behnken experiments. In contrast, it would seem that optimum medium formulation for *C. paraputrificum* biogas production had already been achieved during the Method of Steepest Ascent experiments, since the total volume of biogas did not change significantly during the Box-Behnken experiments 5400 ml/l versus 5350 ml/l from the syringe cultures and 5405 ml/l from the bioreactor batch cultures.

6.5 Conclusion.

Validating this model; according to the mathematical predictions resulting from the Box-Behnken experiments, the maximum predicted total biogas volume were 6478.95 ml/l for *C. beijerinckii* and 5397.78 ml/l for *C. paraputrificum*. The total volume of biogas actually produced during the experiments was 6533 ml/l for *C. beijerinckii* and 5350 ml/l for *C. paraputrificum* from the syringe cultures and 6610 ml/l for *C. beijerinckii* and 5405 ml/l for *C. paraputrificum* from the bioreactor cultures. The correlation between the maximum predicted values and the total biogas volume measured experimentally authenticates the model validation and existence of an optimal independent variable set conditions and

223

this also proves that the optimum formulation to maximize biogas production had been achieved.

Considering the interaction effects for *C. beijerinckii* cultures N-acetylglucosamine concentrations on total biogas volume produced was independent of $FeSO_4.7H_2O$ concentrations and initial pH ranges. The effect of $FeSO_4.7H_2O$ concentrations on total biogas volume was also independent of initial pH ranges. For *C. paraputrificum* cultures interaction effects of N-acetylglucosamine concentrations on the total volume of biogas produced were independent of the L-cysteine.$HCl.H_2O$ and $MgCl_2$ concentrations. The effects of L-cysteine.$HCl.H_2O$ concentrations on the total volume of biogas produced were also independent of the $MgCl_2$ medium concentrations.

Chapter 7

General Discussion and Conclusion

7.1 Discussions and Conclusion.

The formulation of growth media for biogas production by *Clostridium beijerinckii* and *C. paraputrificum* when using N-acetylglucosamine as the main carbon source was performed using rational statistical methods. To the best of this author's knowledge, it has not previously been reported whether *C. beijerinckii* can utilize N-acetylglucosamine as a carbon source; however other workers have shown that it is able to utilize colloidal chitin and that it produces a variety of chitinolytic enzymes (Simunek, Tishchunko & Koppova, 2008). In this study *C. beijerinckii* was, for the first time, found to be capable of metabolising N-acetylglucosamine. On the other hand, *Clostridium paraputrificum* has previously been shown to metabolise both chitin and N-acetylglucosamine, producing hydrogen, CO_2 and a variety of organic acids in the process (Evvyernie et al., 2000). However statistical optimization of the growth medium for biogas production by *C. paraputrificum* was not conducted.

Initially, in the work described in this work, the Plackett and Burman design was used to determine the relative importance, for the maximal production of biogas, of a variety of different medium constituents. For *C. beijerinckii* the initial pH of the culture medium and also the initial $FeSO_4.H_2O$ concentration were identified as significant factors for the production of biogas. On the other hand, for the *C. paraputrificum* cultures the initial N-acetylglucosamine, L-cysteine.$HCl.H_2O$ and $MgCl_2$ concentrations were all identified as significant factors for the production of biogas. Technically, a medium constituent is considered to have had a "real" effect upon a factor of interest when its effect upon that factor is found to be statistically significant; in other words the effect upon the factor did not occur by chance but occurred as a direct result of a change to the medium constituent (Greasham & Herber, 1997).

To demonstrate that the effect of N-acetylglucosamine should not be ignored for *C. beijerinckii* cultures, in the next phase of optimization using the Method of Steepest Ascent, the range of effects (Table 3.15) was analyzed and suggested that there were some interactions between medium constituents concealing the effect of N-acetylglucosamine. According to the results of the Plackett and Burman medium design trials, L-cysteine.HCl.H$_2$O had the 3rd greatest effect upon *C. beijerinckii* biogas production. L-cysteine.HCl.H$_2$O reduces the effect of dissolved oxygen in the culture medium (Sinji et al., 1998). This suggests that oxygen ingress may have been a problem in the experimental procedures. Consequently, to determine the optimum concentration of L-cysteine.HCl.H$_2$O needed to reduce the redox potential of the medium to a suitable level for an anaerobic bacterium such as *C. beijerinckii*, the one factor technique was used at various L-cysteine.HCl.H$_2$O concentrations. Trials with high initial concentrations of L-cysteine.HCl.H$_2$O in the medium demonstrated that the bacteria initiated visible production of biogas sooner than those with lower initial concentrations of it. The lag times before fermentation was initiated were also shorter in cultures with high initial L-cysteine.HCl.H$_2$O concentrations. L-cysteine.HCl.H$_2$O helped to make the environment suitable for anaerobic bacteria by scavenging any dissolved oxygen that might be present in the medium (Sinji, et al., 1998). L-cysteine reduces the redox potential of the culture medium to a level suitable for cell growth of *C. beijerinckii* to be initiated (Sim & Kamaruddin, 2008). Inhibition of clostridial growth by the presence of oxygen is because these species lack the necessary mechanisms to remove oxygen derivatives and reactive by-products such as hydrogen peroxide, hydroxyl radicals and superoxide (Sinji et al., 1998). Bacteria form systems for protection against oxidative stress, using enzymes such as catalase and superoxide dismutase (Cabiscol et al., 2000). The lack of a catalase enzyme is the most significant factor for *Clostridium* aerointolerance (Holland et al., 1987). Interestingly, as the initial cysteine concentration of the medium was increased beyond a critical level there was actually a reduction in the final volume of biogas produced. L-cysteine is highly reactive and is known to be a good chelator of metal ions, and it may be that when present to excess the L-cysteine may combine with components of the nutrient medium. This could result in reducing the bioavailability of some nutrients and lead to a decline in *C. beijerinckii* bioactivity (Yu et al., 2007).

After the optimum L-cysteine.HCl.H_2O concentration required supporting maximal biogas yield, the Method of Steepest Ascent was employed to further optimize the medium for maximal total volume biogas produced. The results indicated that the medium formulation that was close to the optimum for biogas production by *C. beijerinckii* included an initial concentration of 17 g/l of N-acetylglucosamine, 0.08 g/l of $FeSO_4.7H_2O$ and an initial pH of 6.3 (Table 5.1), whilst for *C. paraputrificum* it was initial concentrations of 27 g/l of N-acetylglucosamine, 0.25 g/l of L-cysteine.HCl.H_2O and 0.45 g/l of $MgCl_2$ (Table 5.2).

The Box–Behnken method was used to further refine the optimal level of the individual medium constituents and also to determine any interactions between them. The F-values of 13.97 (*C. beijerinckii*) and 9.18 (*C. paraputrificum*) show that the regression model is significant. The data showed that, for the *C. beijerinckii* cultures, the effect of the initial N-acetylglucosamine concentration on total biogas yield was independent of both the initial $FeSO4.7H_2O$ concentration and the initial pH. Furthermore, the effect of the initial $FeSO_4.7H_2O$ concentration on total biogas yield was also independent of the initial pH. Similarly, for the *C. paraputrificum* cultures the effect of N-acetylglucosamine concentration on the total biogas yield was independent of both the initial L-cysteine.$HCl.H_2O$ and $MgCl_2$ concentrations in the medium. The effect of the initial L-cysteine.$HCl.H_2O$ concentration in the medium on the total biogas yield was also independent of the initial concentration of $MgCl_2$ in the medium. This is because their individual interaction effects were insignificant.

The squared term effect describes the relationship between an investigated medium constituent and the target response (total volume of biogas). In addition, a direct linear relationship between the initial concentration of N-acetylglucosamine in the medium and the total biogas yield, and also the initial pH of the medium and the total biogas yield, was observed for *Clostridium beijerinckii*. In contrast, the relationship between the initial $FeSO4.7H_2O$ concentration and the total biogas yield was not linear but a curved line. On the other hand there was a direct linear relationship between the initial concentration of $MgCl_2$ in the medium and the total biogas yield of *C. paraputrificum* cultures, whilst the relationship between the initial N-acetylglucosamine concentration and total biogas yield, and also the initial L-cysteine.HCl.H$_2$O concentration and total biogas yield both followed a curved line. The data demonstrated that N-acetylglucosamine had a statistically significant effect upon *C. beijerinckii*, indicating that the total volume of biogas produced varies directly with the initial concentration of N-acetylglucosamine available. Neither the initial FeSO$_4$.7H$_2$O concentration nor the initial pH had a significant effect on the total volume of biogas produced for *C. beijerinckii* cultures. For *C. paraputrificum* cultures, N-acetylglucosamine, L-cysteine.HCl.H$_2$O and MgCl$_2$ have significant linear effects on the total volume of biogas produced. The total volume of biogas produced differs depending on the individual concentrations of N-acetylglucosamine, L-cysteine.HCl.H$_2$0 and MgCl$_2$. However these interpretations are described within the boundaries of the medium constituent concentrations used.

The Box-Behnken method was used to design the optimum medium formulation to achieve maximal total biogas yield from both *C. beijerinckii* and *C. paraputrificum* cultures. These formulations were then tested by culturing *C. beijerinckii* and *C. paraputrificum* in 50 ml syringes, and also in batch cultures of 1 litre, to compare the actual yield of biogas with the predicted values. During all of these fermentations N-acetylglucosamine was converted into a combination of biomass, biogas, energy and organic acids. It was found that the culture medium pH decreased significantly as the N-acetylglucosamine supply in the medium was consumed, and this change in pH correlated with the accumulation of organic acids in the medium. It has previously been observed that changes to the pH of the culture medium can influence both the yield of the final product(s) (Zhu & Yang, 2004) and the range of organic acids produced can vary as a result of changes in the metabolic pathways used by microorganisms (Evvyernie et al., 2000).

In addition to affecting the acid production profile, high concentrations of organic acids in the culture medium, and low pH of the culture medium, can inhibit bio-hydrogen yields (Van Ginkel & Logan, 2005; Khanal et al., 2004). It is also known that low culture medium pH can lead to reduced intracellular availability of ATP which can then cause inhibition of bacterial cell growth and further loss of metabolic activity (Bowles & Ellefson, 1985).

In contrast bio-hydrogen production is aided by the intracellular hydrogenase enzyme. The intracellular hydrogenase is an iron-containing Fe-hydrogenase that catalyzes the oxidation of molecular hydrogen from protons and electrons (Nicolet et al., 2002). Consequently, the concentration of Fe^{2+} in the growth medium can influence bio-hydrogen production (Wang and Wan, 2008). The results of the current experiments indicated that the initial Fe^{2+} concentration of the growth medium did indeed have a significant effect upon biogas yield, in the *C. beijerinckii* cultures. However, up to a critical concentration, increases in the Fe^{2+} concentration led to increases in biogas yield, but above the critical value further increase in the Fe^{2+} concentration actually led to decreases in biogas yield.

In addition to the requirement for Fe^{2+}, Mg^{2+} can also be required in the growth medium, with Mg^{2+} previously having been shown to influence both bacterial cell growth and production of bio hydrogen by *C. beijerinckii* (Zhao et al., 2012).

231

Interestingly, in the current experiment Mg^{2+} was identified to be a significant medium constituent required by *C. paraputrificum* cultures but not by *C. beijerinckii* cultures in the Plackett and Burman experiment. Iron and Magnesium are both important, as Fe^{2+} is a co-factor for the hydrogenase enzyme whilst Mg^{2+} ions are involved in many enzyme activities and also function to stabilize cell membrane (Zhao et al., 2012).

When biogas produced from N-acetylglucosamine as the sole carbon source by the *C. beijerinckii* and *C. paraputrificum* cultures was compared, it can be concluded that *C. beijerinckii* produced a greater total volume of biogas at 6610 ml of gas/l of medium compared with *C. paraputrificum* which produced 5405 ml of gas/l of medium (Table 7.1). However, the calculated yield of bio-hydrogen by *C. paraputrificum* from N-acetylglucosamine at 2.55 mol of H_2 / mol of N-acetylglucosamine was greater than that by *C. beijerinckii* at 2.43 mol of H_2 / mol of N-acetylglucosamine (Table 7.1). Other workers have recorded maximum hydrogen yields of 1.96 mol H_2/ mol of glucose (Zhao et al., 2012), 1.97 mol H_2/ mol of glucose (Zhao et al., 2011), 2.2 mol H_2/ mol of glucose (Guo et al., 2009), 2.2 mol H_2/ mol of glucose (Chong et al., 2009), 2.53 mol H_2/ mol of glucose (Pan et al., 2008), 3.10 mol H_2 / mol of sucrose (Yaun et al., 2008), 1.9 mol H_2/ mol of N-acetylglucosamine (Evvyernie et al., 2000) and 2.2 mol H_2/ mol of N-acetylglucosamine (Evvyernie et al., 2001), most of which are slightly lower than those recorded in the current study and suggest that N-acetylglucosamine might be a good carbon source for bio-hydrogen production.

However comparing the results in this study when N-acetylglucosamine was utilized, to results reported by Yaun et al. (2008) with a maximum bio-hydrogen yield of 3.10 mol H_2 / mol of sucrose, a higher bio-hydrogen yield from sucrose was recorded, and this may be a result of sucrose being a disaccharide and containing more carbon and hydrogen per mole than N-acetylglucosamine. Nevertheless N-acetylglucosamine would likely have greater commercial value as a carbon source for bioenergy production as it is a waste product derived from chitin and not a primary food resource. The research of Evvyernie et al. (2000) with a maximum bio-hydrogen yield of 1.9 mol H_2/ mol of N-acetylglucosamine and Evvyernie et al. (2001) with maximum bio-hydrogen yield of 2.2 mol H_2/ mol of N-acetylglucosamine had much lower yields than this study. However both of these previous studies did not statistically optimize hydrogen production. Therefore this is the first study, to the author's knowledge, to statistically optimize total biogas produced from N-acetylglucosamine and this resulted in significant increases in biogas production. The yields obtained were greater than or at least equivalent to the yields previously achieved by other authors for biogas production from glucose and as a result demonstrate significant potential for the use as chitin as an alternative feedstock for commercial biogas production from waste products. A greater bio-hydrogen yield was achieved by the statistical optimization techniques in this work. A summary for the biogas optimization process is shown in figure 7.1

Clostridium beijerinckii

Plackett and burman method
- Identifies significant medium constituents variables
- Initial pH and $FeSO_4.7H_2O$ were significant
- Biogas improved from **1900 ml/l to 3183 ml/l.**

Cysteine one factor optimization
- Used to confirm there was no oxygen ingress affecting the optimization technique.
- Biogas improved from **3183 ml/l to 5467 ml/l**

Method of Steepest Ascent
- GlcNAG, initial pH and $FeSO_4.7H_2O$ were optimized further to the vicinity of optimum.
- Biogas improved from **5467 ml/l to 5733 ml/l**

Box-Behnken method
- Identifies interaction effect amongst medium constituents
- Identifies the optimum medium formulation for maximum biogas volume.
- Biogas improved to a maximum from **5733ml/l to 6610 ml/l.**

Clostridium paraputrificum

Plackett and Burman method
- Identifies significant medium constituents variables
- GlcNAG, $MgCl_2$ and $FeSO_4.7H_2O$ were significant
- Biogas improved from **0 ml/l to 1300 ml/l.**

Method of Steepest Ascent
- GlcNAG, $MgCl_2$ and $FeSO_4.7H_2O$ were optimized further to the vicinity of optimum.
- Biogas improved from **1300 ml/l to 5400 ml/l**

Box-Behnken method
- Identifies interaction effect amongst medium constituents
- Identifies the optimum medium formulation for maximum biogas volume.
- Biogas improved to a maximum from **5400 ml/l to 5405 ml/l.**

Figure 7.1 Decision tree-type flow diagrams showing the process of optimization and increases in cumulative biogas volume

Table 7.1 Yield coefficient of moles of biogas per mole of N-acetylglucosamine consumed, at stages of the statistical optimization methods of *C. beijerinckii* and *C. paraputrificum* at optimum conditions using the ideal gas law assuming conditions of temperature at 37°C and pressure at 1 atm.

Optimisation Method	*Clostridium* Trials	Biogas Volume (ml/l)	Biogas moles	Moles of biogas/ mole of GlcNAc	Moles of CO_2 / mole GlcNAc	Moles of H_2 / mole GlcNAc
Plackett and Burman	*C. beijerinckii*	3183	0.125	1.814	0.082	1.732
	C. paraputrificum	1300	0.051	0.863	0.190	0.672
Method of Steepest Ascent	*C. beijerinckii*	5733	0.225	3.153	0.364	2.789
	C. paraputrificum	5400	0.212	2.090	0.298	1.792
Box-Behnken	*C. beijerinckii*	6610	0.260	2.851	0.418	2.433
	C. paraputrificum	5405	0.212	3.037	0.491	2.546

ml/l = millilitres of biogas per litre of medium.

The results of the statistical optimization using Plackett and Burman design, the Method of Steepest Ascent and the Box-behnken design used to screen key variables and identify optimal values of medium constituents for biogas production in this study have significantly increased the hydrogen yields.

The optimization procedure was carried out sequentially by first identifying the key significant medium constituents that influence the production of maximum cumulative biogas volume. These significant constituents were not known in the initial medium formulation. These key medium constituents were further explored, studied and there concentrations manipulated by the method of steepest ascent as well as the Box-behnken method to further identify medium constituent interactions on cumulative biogas volume. The information derived from this was further explored to identify the optimum final medium formulation for maximum cumulative biogas volume compared to the initial medium used.

Therefore the statistical design techniques were useful in optimizing the bio-hydrogen yield in *C. beijerinckii* and *C. paraputrificum* form utilizing N-acetylglucosamine. Finally, the authenticity of the statistical optimization method was confirmed, there was good consistency between the predicted biogas total volume by the clostridial species and the actual total biogas volume produced, when they were supplied with N-acetylglucosamine as the sole carbon source in the optimized medium formulations.

Future Work

Future work

Clostridium beijerinckii and *C. paraputrificum* produced substantial quantities of organic acids as end products during the production of biogas. The production of these organic acids corresponded with the culture pH decline, in turn leading to a cease in utilization of N-acetylglucosamine. To prove that low culture pH is responsible for the cease in utilization of N-acetylglcosamine for biogas production, in the future work the culture pH will be maintained at pH 6.5 in a continuous culture.

The large quantities of organic acids produced during the utilization of N-acetylglucosamine contain large amounts of carbon energy. The effective utilization of these stored carbon energy will also be investigated. *Clostridium beijerinckii* or *C. paraputrificum* and a photo-fermentative bacterium such as *Rhodopseudomonas faecalis* will be employed to produce biogas in mixed cultures utilizing N-acetylglucosamine as the sole substrate. Due to the difference in growth rate, the demand for light energy and organic acid tolerance between the dark-fermentative bacteria and the photo-fermentative bacteria, biogas yield, organic acids, culture pH and biomass at different time intervals will be investigated. Organic acids produced by *C. beijerinckii* and *C. paraputrificum* will serve as effective substrates for the photo-fermentative bacteria. This combination will form a food chain and ensure the effective utilization of N-acetylglucosamine for biogas production.

Bibliography

Bibliography

Abdel-Fattah, Y.R. & Olama, Z.A. (2002). Asparaginase production by *Pseudomonas aeruginosa* in solid-state culture: Evaluation and optimization of culture conditions using factorial designs. *Process Biochemistry*, 38, 115–122.

Alalayah, W.M., Mohd, S. K., Abdul A.H., Kadhum, J.M. Jahim & Najeeb M. A. (2009). Effect of Environmental Parameters on hydrogen Production using *Clostridium Saccharoperbutylacetonicum* N1-4 (ATCC 13564). *American Journal of Environmental Sciences, 5*(1), 80-86.

Alshiyab, H., Kalil. S.M., Hamid, A. & Yusoff, W. (2008). Trace Metal Effect on Hydrogen Production Using *Clostridium acetobutylicum*. *Journal of Biological Sciences 8*(1): 1-9.

Antony, J. (2006). Taguchi or classical design of experiments: A perspective from a practitioner. *Sensor review*, 26, 227–230.

Argun, H., Kargi, F., Kapdan, I., Oztekin, R. (2008). Biohydrogen production by dark fermentation of wheat powder solution: Effects of C/N and C/P ratio on hydrogen yield and formation rate. *International Journal of Hydrogen Energy*, 33, 1813–1819.

Bakonyi, P., Nemesto thy, N., Lo vitusz E . & Be lafi-Bako, K. (2011). Application of Plackett-Burman experimental design to optimize biohydrogen fermentation by E. coli (XL1-BLUE). *International Journal of Hydrogen Energy, 36*(21), 13949-13954.

Bartram J. & Balance, R. (1996). *Water Quality Monitoring - A Practical Guide to the Design and Implementation of Freshwater Quality Studies and*

Monitoring Programmes. London: E&FN Spon.

Benemann, J.R. (1997). Feasibility analysis of photobiological hydrogen production. *Int J Hydrogen Energy*, 22, 979- 987.

Benemann, J.R., Berenson, J.A., Kaplan, N.O. & Kauren, M.D. (1973). Hydrogen evolution by a chloroplast-ferredoxin-hydrogenase system. *Proc Natl Acad Sci*, 70, 2317-2320.

Bisaillon. A., Turcot, J., Hallenbeck, P. (2006). The effect of nutrient limitation on hydrogen production by batch cultures of *Escherichia coli*. *International Journal Hydrogen Energy*, 31, 1504–1508.

Blackwood, A.C., Neish, A.C. & Ledingham, G.A. (1956). Dissimilation of glucose at controlled pH values by pigmented and non-pigmented strains of *Escherichia coli*. *Journal of Bacteriology, 72*, 497-499.

Bose, T., Malbrunot, P., Benard, P. & Viola, J. (2007). *Hydrogen: Facing the energy challenges of the 21st century*. Esher: John Libbey Eurotext.

Bowles, L.K. & Ellefson, W.L. (1985) Effects of Butanol on *Clostridium acetobutylicum. Applied and Environmental Microbiology, 50*(5): 1165-1170.

Box, G. E. P. & Wilson, K. B. (1951). On the experimental attainment of optimum conditions. *Journal of the Royal Statistical Society Series B, 13*(1), 1-45.

Cabiscol, E., Tamarit, J., Ros, J., (2000) Oxidative stress in bacteria and protein damage by reactive oxygen species. *International Microbiology, 3*:3–8.

Cai, G., Jin, B., Saint, C., & Monis, P. (2010). Metabolic flux analysis of hydrogen production network by Clostridium butyricum W5: Effect of pH and glucose concentrations. *International journal of hydrogen energy 35*(13),

1-10

Cammack, R. (1999). Hydrogenase sophistication, *Nature,* 397, 214–215.

Cammack, R., Frey, M. & Robson, R. (2001). *Hydrogen as a fuel: learning from Nature.* London: Taylor and Francis.

Canganella, F., Kuk, S.U., Morgan, H. & Wiegel, J. (2002). *Clostridium thermobutyricum*: growth studies and stimulation of butyrate formation by acetate supplementation. *Microbiological Research.* 157(2), 149-156.

Carlsson, J., Granberg, G., Nyberg, G. & Edlund. M. (1979). Bactericidal Effect of cysteine exposed to atmospheric oxygen. *Applied and environmental microbiology*; *37*(3), 383-390.

Chen, C.C, Lin, C.Y., Chang, J.S. (2001). Kinetics of hydrogen production with continuous anaerobic cultures utilizing sucrose as the limiting substrate. *Applied Microbiology and Biotechnology,* 57, 56-64.

Chenlin Li and Herbert H. (2007). Fermentative Hydrogen Production From Wastewater and Solid Wastes by Mixed Cultures. *Environmental Science and Technology*, 37, 1–39.

Chittibabu, G., Nath, K. & Das, D. (2006). Feasibility studies on the fermentative hydrogen production by recombinant *Escherichia coli* BL-21. *Process Biochemistry*, *41*, 682–688.

Chong, M., Nor'Aini, A., Phang, L.Y., Suraini, A., Raha, A., Yoshihito, S., Mohd, A., (2009). Effects of pH, Glucose and Iron Sulphate Concentration on the yield of biohydrogen by Clostridium butyricum EB6. International journal of hydrogen energy, 34, 8859 – 8865.

Chou, C.H., Wang, C.W., Huang, C.C., Lay, J.J. (2008). Pilot study of the influence of

242

stirring and pH on anaerobes converting high- solid organic wastes to hydrogen. *International Journal of Hydrogen Energy, 33,*1550–1558.

Chuannan, L., Jingjing C., Zuotao, L., Yuntao, L., Minnan, L. & Zhong, H. (2010). Statistical optimization of fermentative hydrogen production from xylose by newly isolated Enterobacter sp. CN1. *International Journal of Hydrogen Energy. 35*(13),6657-6664.

Chun-Mei, P., Yao-Ting, F., Pan, Z. & Hong-Wei, H. (2008). Fermentative hydrogen production by the newly isolated *Clostridium beijerinckii* Fanp3. *International Journal of Hydrogen Energy. 33,(20)*, 5383-5391.

Claassen P.A.M., Budde M.A.W., van Niel E.W.J and de Vrije T. (2005). Biofuels for fuel cells. Renewable energy from biomass fermentation. In D. Piet, L., Peter, Westermann., Marianne, H. and Angelo, M. (Eds.), *Utilization of biomass for hydrogen fermentation.* (pp). London: IWA.

Claassen, P. A. M., van Lier, J. B., Contreras, A. M. L., van Niel, E. W. J., Sijtsma L., Stams, A. J. M., de Vries, S. S. & Weusthuis R.A. (1999). Utilisation of biomass for the supply of energy carriers. *Applied. Microbiology. Biotechnology. 52,* 741-755.

Dabrock, B., Bahl, H. & Gottschalk, G. (1992). Parameters affecting solvent production by *Clostridium pasteurianum. Applied Environmental Microbiology, 58,* 1233–1239.

Das, D. and Veziroˉglu, T.N. (2001). Hydrogen production by biological processes: A survey of literature. *Int. J. Hydrogen Energy, 26*(1), 13–28.

De Vrije, T. and Claassen, P. A. M. (2003). Dark hydrogen fermentations. In Reith, J.H., Wijffels, R.H. & Barten, H. (Eds), *Bio-methane & Bio-hydrogen:*

Status and Perspective of biological methane and hydrogen production. (pp. 103-121). Netherland ECN: Dutch biological hydrogen foundation.

Dietrich, G., Weiss, N. & Winter, J. (1988). *Acetothermus paucivorans*, gen. Nov., sp. nov, a strictly anaerobic, thermophilic bacterium from sewage sludge, fermenting hexoses to acetate, CO_2 and H_2. *System Applied Microbiology, 10,* 174-179.

Ding, J., Liu, B., Ren, N., Xing, D., Guo, W., Xu, J., Xie, G. (2009). Hydrogen production from glucose by co-culture of *Clostridium Butyricum* and immobilized *Rhodopseudomonas faecalis RLD-53. International journal of hydrogen energy,* 34, 3647–3652.

Donderski, W., Swiontek, B. M. (2003). The Utilization of N-acetyloglucosamine and Chitin as Sources of Carbon and Nitrogen by Planktonic and Benthic Bacteria in Lake Jeziorak. *Polish Journal of Environmental Studies,12*(6),*685-692.*

Duncan, S.H., Holtrop, G., Lobley, G.E., Calder, A.G., Stewart, C.S., Flint, H.J. (2004). Contribution of acetate to butyrate formation by human faecal bacteria. *British Journal of Nutrition, 91,* 915–923.

Espinoza-Escalante, F.M., Pelayo-Ortiz, C., Gutie´ rrez-Pulido, H., Gonzalez-A´ lvarez, V., Alcaraz-Gonza´ lez, V., Bories, A. (2008). Multiple response optimization analysis for pretreatments of Tequila's stillages for Volatile fatty acids and hydrogen production. *Bioresource Technology,*99, 5822–5829.

European commission. (2002). *New carriers and novel technologies for a future clean and sustainable energy economy.* Luxembourg: Office for official publications of the European communities.

Evvyernie, D., Morimoto, K., Karita, S., Kimura, T., Sakka, K., Ohmiya, K. (2001). Conversion of chitinous wastes to hydrogen gas by *clostridium paraputrificum* M-21. *Journal of Bioscience and Bioengineering,91*(4), 339-343.

Evvyernie, D., Yamazaki, S., Morimoto, K., Karita, S., Kimura, T., Sakka, K., Ohmiya, K. (2000). Identification and Characterization of Clostridium paraputrificum M-21, a Chitinolytic, Mesophilic and Hydrogen-Producing Bacterium. *Journal Of Bioscience and Bioengineering, 89*(6), 596-601.

Fabiano, B. & Perego, P. (2002). Thermodynamic study and optimization of hydrogen production by *Enterobacter aerogenes. International Journal Hydrogen Energy*, 27,149-156.

Fan, Y.T. & Zhang, Y.H. (2006). Efficient conversion of wheat straw wastes into bio hydrogen gas by cow dung compost. *Bioresour. Technology*; *97*, 500–505.

Fan, Y.T., Li, C.L., Lay, J.J., Hou, H.W. & Zhang, G.S. (2004). Optimization of initial substrate and pH levels for germination of sporing hydrogen-producing anaerobes in cow dung compost. *Bioresour. Technology, 91*, 189–193.

Fang, H.P. & Hong, L. (2002). Effect of pH on hydrogen production from glucose by a mixed culture. *Bioresource Technology, 82*(1), 87–93.

Fumiaki, T., Jun Dan h., Shuya, T. & Masayoshi, M., (1992). Efficient hydrogen production from starch by a bacterium isolated from termites. *Journal of Fermentation and Bioengineering, 73*(3), 244-245.

Godfroy, A., Raven, N.D. & Sharp, R.J. (2000). Physiology and continuous culture of the hyperthermophilic deep-sea vent archaeon *Pyrococcus abyssi*

ST549. *FEMS Microbiology Letters, 186*, 127-132.

Gosselink, J.W. (2002). Pathways to a more sustainable production of energy: sustainable hydrogen - a research objective for Shell. *International Journal of Hydrogen Energy, 27*, 1125 – 1129.

Gray, C.T. & Gest, H. (1965). Biological formation of molecular hydrogen. *Science, 148*, 186-192.

Greasham, R. & Herber, W. (1997). Design and optimization of growth media. In P. Rhodes, & P.Stanbury (Eds.), *Applied microbial physiology*: A practical approach. (pp. 53-73). New york: Oxford university press.

Greenwood, D. & Butcher, M. (1997). Bringing the issue of documentation to the ward staff. In D. Leaper, C. Dealey, P.J. Franks, D. Hofman & C. Moffatt (Eds.), *Proceedings of the 7th European Conference on Advances in Wound Management.* (pp.134-136). Cambridge: University Press.

Guo, W.Q., Ren, N.Q., Wang, X.J., Xiang, W.S., Ding, J. & You, Y. (2009). Optimization of culture conditions for hydrogen production by *Ethanoligenens harbinense* B49 using response surface methodology. *Bioresource Technology, 100*, 1192–1196.

Hamelinck, C. N., Faaij, A. P. C., Uil, H. D. & Boerrigter, H. (2003). Production of FT transportation fuels from biomass; technical options, process analysis and optimisation, and development potential. *Energy, 29*(11), 1743-1771.

Hartmanis, M., Klason, T. & Gatenbeck, S. (1984). Uptake and activation of acetate and butyrate in *Clostridium acetobutylicum. Applied Microbiology and Biotechnology,* 20(1), 66-71.

Hartmanis, M.N. & Gatenbeck, S. (1984). Intermediary Metabolism in *Clostridium acetobutylicum*: Levels of Enzymes Involved in the Formation of Acetate

246

and Butyrate. Applied and Environmental Microbiology, 47(6), 1277-1283.

Hawkes, F.R., Dinsdale, R., Hawkes, D.L. & Hussy, I. (2002). Sustainable fermentative biohydrogen: challenges for process optimization. *International Journal of Hydrogen Energy, 27,* 1339–1347.

He, G., Kong, Q., Chen, Q., Ruan, H., (2005). Batch and fed-batch production of butyric acid by *Clostridium butyricum* ZJUCB. *Journal of Zhejiang University ,6B*(11),1076-1080.

Heyndrickx, M., Vansteenbeeck, A., Vos, P. & Ley, L. (1986). Hydrogen gas production from continuous fermentation of glucose in a minimal medium with *Clostridium butyricum* LMG 1213tl. *System Applied Microbiology*, 8, 239-244.

Heyndrickx, M., Vos, P. & Ley, L. (1990). Hydrogen production from chemostat fermentation of glucose by *Clostridium butyricum* & *Clostridium pasterianum* in ammonium- and phosphate limitation. *Biotechnology Letters, 12,* 731-736.

Hoffmann, P. (2004). Tomorrow's Energy: Hydrogen, *Fuel Cells and the Prospects for a Cleaner Planet.* Cambridge: MIT Press.

Holland, K. T., Knapp, J. S., and Shoesmith, J. G (1987). Anaerobic bacteria. The Chapman and Hall, New York.

Hsu, S.T., Yang, S.T. (1991). Propionic acid fermentation of lactose by *Propionibacterium acidipropionici*: effects of pH. *Biotechnology Bioenginering*, 38, 572–578.

Huges, M.N. & Poole, R.K. (1989). Metals and Micro-organisms. London: Chapman and Hall.

Hwang, M.H., Jang, N.J., Hyun, S.H., and Kim, I.S. (2004). Anaerobic bio-hydrogen production from ethanol fermentation: The role of pH, *J. Biotechnol. 111*(3), 297–309.

Ida, A., Marcel, J., Jorge, R. & Renee H. (2002). Photobiological hydrogen production: photochemical efficiency and bioreactor design. *International Journal of Hydrogen Energy* ,27,1195 – 1208.

Ilgi, K.K. and Fikret K. (2006). Bio-hydrogen production from waste materials. *Enzyme and Microbial technology*, 38, 569-582.

Innotti, E.L., Kafkawitz, D., Wolin, M.J. & Bryant, M.P. (1973), Glucose fermentation products of *Ruminococcus albus* grown in continuous culture with *Vibrio succinogenes:* changes caused by interspecies transfer of Hydrogen. *Journal of Bacteriology, 114*, 1231-1240.

Ishikawa, M., Yamamura, S., Takamura, Y., Sode, K., Tamiya, E., Tomiyama, M. (2006). Development of a compact high-density microbial hydrogen reactor for portable bio-fuel cell system. *International Journal Hydrogen Energy*, 31, 1484–1489.

Jaros, A., Rova, U. & Kris. A. Effect of acetate on fermentation production of butyrate. *Cellulose Chemistry and Technology.46*(5), 341-347.

Jiao, A., Yue, Y. & Yansg, C. (2010). Start Up of biohydrogen production system and effect of metal ions on hydrogen production. *American Institute of Physics Conference Proceedings.* (pp. 197-200). Beijing: American Institute of Physics.

Jo, H., Lee, D., Kim, J., Park, P. (2009). Effect of Initial Glucose Concentrations on

Carbon and Energy Balances in Hydrogen-Producing Clostridium tyrobutyricum JM1. *Journal of Microbiology and Biotechnology*, *19*(3), 291–298.

Jo, J. H., Lee,D.S., Donghee P. & Jong M.P. (2008). Statistical optimization of key process variables for enhanced hydrogen production by newly isolated *Clostridium tyrobutyricum* JM1. *International Journal of Hydrogen Energy*, 33, 5176–5183

Jolles, P. & Muzzarelli, R. A. (1999). *Chitin and chitinase.* Berlin: Birkhauser.

Jolles, P., Muzzarelli, R.A.A. (1999). Chitin and Chitinases. In D. Springer, H. M., Wise, D.L., Moo-Young, M. (Eds.*), Environmental Monitoring and Biodiagnostics of Hazardous Contaminants.*(pp 352). New York: Springer.

Jones, A. & Pickard, M. D., (1980). Effect of Titanium (III) Citrate as Reducing Agent on Growth of Rumen Bacteria. *Applied and Environmental Microbiology*, *39*(6), 1144-1147.

Joon, L.Y., Takashi, M., Noike, T., (2001). Effect of iron concentration on hydrogen fermentation. *Bioresource Technology,* 80, 227-231.

Jung, G.Y., Kim, J.R., Park, J.Y. & Park, S. (2002). Hydrogen production by a new chemoheterotrophic bacterium *Citrobacter sp.* Y19. *International Journal of Hydrogen Energy, 27,* 601-610.

Kalia, V.C., Jain, S.R., Kumar, A. & Joshi, A.P. (1994). Fermentation of bio waste to H_2 by *Bacillus licheniformis. World Journal of Microbiology and Biotechnology, 10,* 224-227.

Kapdan, I. K. and Kargi. F. (2006). Bio-hydrogen production from waste materials. *Enzyme and Microbial Technology*, 38, 569- 582.

Kari, C., Nagy, Z., Kovacs, P. & Hernadi. F. (1971). Mechanism of the growth inhibitory effect of cysteine on Escherichia coli. *J. Gen. Microbiol*, *68*, 349-356.

Karthic, P., Joseph, S. & Arun, N. (2012). Optimization of Process Variables for Biohydrogen Production from Glucose by Enterobacter aerogenes. *Scientific Reports*, *1*(2), 173.

Kawasaki, S., Nakagawa, T., Nishiyama, Y., Benno, Y., Uchimura, T., Komagata, K., Kozaki, M. and Niimura, Y. (1998). Effect of oxygen on the growth of *Clostridium butyricum* (type species of the genus *Clostridium*), and the distribution of enzymes for oxygen and for active oxygen species in Clostridia. *Journal of Fermentation and Bioengineering, 86*(4), 368-372.

kawasaki, S., Nakagawa, T., Nishiyama, Y., Benno, Y., Uchimura, T., Komagata, K., Kozaki, M., & Niimura, Y., (1998). Effect of oxygen on the growth of *Clostridium butyricum* (type species of the genus *Clostridium*), and the distribution of enzymes for oxygen and for active oxygen species in clostridia. *Journal Of Fermentation and Bioengineering, 86*(4), 368-372.

Kennedy, M. & Krouse, D. (1999). Strategies for improving fermentation medium performance: a review. *Journal of Industrial Microbiology & Biotechnology*, *23*, 456–475.

Kennedy, M. J., Reader, S. L., Davies, R. J., Rhoades, D. A & Silby, H. W. (1994). The scale up of mycelial shake flask fermentations: a case study of gamma-linolenic acid production by *Mucor hiemalis* IRL 51. *Journal of Industrial Microbiology*, 13, 212–216.

Khanal, S.K., Chen, W.H., Li, L. & Sung, S.W. (2004). Biological hydrogen production: effects of pH and intermediate products. *International Journal Hydrogen Energy*,29, 1123–1131.

Khor, E. (2001). *Chitin: Fulfilling a Biomaterials Promise.* London: Elsevier.

Kim, S. H., Han, S. K. & Shin, H. S. (2006). Effect of substrate concentration on hydrogen production and 16S rDNA-based analysis of the microbial community in a continuous fermenter. *Process Biochemistry, 41*, 199–207.

Koesnandar, A., Nishio, N. & Nagai, S. (1990). Stimulation by cysteine on growth of *Clostridium thermoaceticum* in minimal medium. *Applied Microbiology and Biotechnology, 32*(6), 711-714.

Koku, H. & Eroğlu, I. (2002). Aspects of metabolism of hydrogen production by *Rhodobacter sphaeroides. Int. J. Hydrogen Energy*, 27, 1315–1329.

Kosaric, N. & Velikonja, J. (1995). Liquid and gaseous fuels from biotechnology: challenge and opportunities. *Federation of European Microbiological Society Microbiology Reviews, 16*, 111-142.

Kruse, O., Rupprecht, J.R., Mussgnug, G.C., Dismukes and Hankamer, B. (2005). *Photochemical & Photobiological Sciences, 4*, 957.

Kuehl, R.O. (2000). *Design of experiments: statistical principles of research design and analysis.* Pacific Grove: Duxbury Press.

Kuhn, M., Steinbuchel, A. & Schlegel, H.G. (1984). Hydrogen evolution by strictly aerobic Hydrogen bacteria under anaerobic condition. *Journal of Bacteriology, 159*, 633-639.

Kumar, A., Jain, S. R., Sharma, C. B., Joshi, A. P. & Kalia, V. C. (1995). Increased hydrogen production by immobilized microorganisms. *World Journal of Microbiology Biotechnology, 11*, 156-159.

251

Kumar, N., Das, D. (2000). Enhancement of hydrogen production by Enterobacter cloacae IIT-BT 08. *Process Biochemistry, 35*, 589-593.

Kumar, N., Ghosh, A. & Das, D. (2001). Redirection of biochemical pathways for the enhancement of H_2 production by *Enterobacter cloacae. Biotechnology Letters*, 23, 537-541.

Lay, J. (2000). Modeling and optimization of anaerobic digested sludge converting starch to hydrogen. *Biotechnology Bioengineering, 68*, 269 – 278.

Lay, J. (2001). Biohydrogen generation by mesophilic anaerobic fermentation of microcrystalline cellulose. *Biotechnology Bioengineering, 74*, 280 – 287.

Lee, D.Y., Li, Y.Y., Oh, Y.K., Kim, M.S. & Noike, T., (2009). Effect of iron concentration on continuous H2 production using membrane bioreactor. *International Journal Hydrogen Energy*, 34:1244-1252.

Lee, Y., Miyahara, J., Noike, T., (2001). Effect of iron concentration on hydrogen fermentation. *Bioresource Technology*, 80, 227-231.

Levin, D., Islam, R., Cicek, N. & Sparling, R. (2006). Hydrogen production by *Clostridium thermocellum* 27405 from cellulosic biomass substrates. *International Journal of Hydrogen Energy*, 31, 1496–1503.

Li, C. L. & Fang, H. H. P. (2007). Fermentative hydrogen production from wastewater and solid wastes by mixed cultures. *Critical Reviews in Environmental Science and Technology, 37*(1), 1–39.

Li, C., Bai, J.H., Cai, Z.L. & Ouyang, F. (2002). Optimization of a cultural medium for bacteriocin production by *Lactococcus lactis* using response surface methodology. *J. Biotechnology, 93*, 27–34.

Li, C.L., Fang, H.H. (2007). Fermentative hydrogen production from wastewater and solid wastes by mixed cultures. *Critical Reviews in Environmental Science and Technology, 37*,1–39.

Lin, C.Y. & Lay, C.H. (2005). A nutrient formulation for fermentative hydrogen production using anaerobic sewage sludge microflora. *International Journal of Hydrogen Energy, 30,* 285–292.

Lin, C.Y. and Lay, C.H. (2004). Carbon/nitrogen-ratio effect on fermentative hydrogen production by mixed microflora. *International Journal of Hydrogen Energy*, 29, 41–45.

Lin, P.Y., Whang, L.M., Wu, Y.R., Ren, W.J., Hsiao, C.J. & Li, S.L. (2007). Biological hydrogen production of the genus Clostridium: metabolic study and mathematical model simulation. *International Journal Hydrogen Energy*, 32:1728–1735.

Liu, B., Ren, N., Tang, J., Ding, J., Liu, W., Xu, J., Cao, G., Guo, W. & Xie, G. (2010). Bio-hydrogen production by mixed culture of photo- and darl-fermentation bacteria. *International journal of hydrogen energy*, 35, 2858-2862.

Liu, G.Q. & Wang, X.L. (2007). Optimization of critical medium components using response surface methodology for biomass and extracellular polysaccharide production by *Agaricus blazei. Appl Microbiol Biotechnology, 74,* 78–83.

Long, C., Cui, J., Liu, Z., Liu, Y., Long, M., & Hu, Z. (2010). Statistical optimization of fermentative hydrogen production from xylose by newly isolated

Enterobacter sp. CN1. *International Journal of Hydrogen Energy, 35*(13), 6657–6664.

Luftig, J. T. & Jordan, V. S. (1998). *Design of experiments in quality engineering.* New York: McGraw-Hill.

Luxembourg. European commission community research. (2002). *Hydrogen and electricity. New carriers and novel technologies for a future clean and sustainable energy economy.* Luxembourg: European commission.

Mandelik, B.G., Newsome, D. and Kellogg, P. (1985). *Kirk-othmer concise encyclopedia of chemical technology.* In D. Grayson, M. (Eds.). (pp 621-623). New York: John Wiley & Sons.

Manikandan, R., Pratheeba, C.N., Pankaj, S. & Stuti, S. (2010). Optimization of Asparaginase production by pseudomonas aeruginosa using Experimental methods. *Nature and science, 8*(2),1-6.

Martin, J.F. (1989). *In regulation of secondary metabolism in Actinomycetes.* CRC Press: Boca Raton.

Masset, J., Hiligsmann, S., Hamilton, C., Beckers, L., Franck, F., Thonart, P. (2010). Effect of pH on glucose and starch fermentation in batch and sequenced-batch mode with a recently isolated strain of hydrogen-producing Clostridium butyricum CWBI1009. *Interational Journal o f Hydrogen Energy, 35,* 3371 – 3378.

Mei-Ling, C., Nor'Aini, A., Phang, L.Y., Suraini, A., Raha, A., Yoshihito, S., Mohd, A., (2009). Effects of pH, Glucose and Iron Sulphate Concentration on the yield of biohydrogen by Clostridium butyricum EB6. International journal of hydrogen energy, 34, 8859 – 8865.

Minguo, Z., Sheng, W. (1978). Effect of metal ions on growth and sporulation of

Clostridium perfringens in a synthetic medium. *Chinese Journal of Microbiology, 11*(2), 50-61.

Minton, P. & Clarke, J. (1989). *Biotechnology handbooks 3. Clostridia*. New York: Springer

Mizuno, O., Dinsdale, R., Hawkes, F.R., Hawkes, D.L. & Noike, T. (2000). Enhancement of hydrogen production from glucose by nitrogen gas sparging. *Bioresource Technology 73*, 59-65.

Montgomery, D.C. (2005). *Design and analysis of experiments*. New York: John Wiley & Sons.

Nandi, R. & Sengupta, S., (1998). Microbial Production of Hydrogen: An Overview. *Critical Reviews in Microbiology, 24*(1), 61–84.

Nandi, R., Sengupta, S. (1996). Involvement of anaerobic reductases in the spontaneous lysis of formate by immobilized cells of *Escherichia coli*. *Enzyme Microbiology Technology, 19*, 20-25.

Nath, K., Das, D. (2004). Improvement of fermentative hydrogen production: various approaches. *Applied Microbiology Biotechnology*, 65, 520–529.

Nemat, O. keyhani and Saul, R, (1996). The chitin catabolic cascade in the marine bacterium *Vibrio furnissii*: Molecular cloning, isolation, and characterization of a periplasmic chitodextrinase. *The journal of biological chemistry, 271* (52), 33414–33424.

Nicolet, Y., Cavazza, C. & Fontecilla-Camps. J.C. (2002). Fe-only hydrogenases: structure, future and evolution. *Journal of Inorganic Biochemistry, 91*, 1–8.

Noike, T. & Mizuno, O. (2000). Hydrogen fermentation of organic municipal waste. *Water Science Technology 42,* 155-162.

255

O'brien, B. and Morris, J. (1971). Oxygen and the growth and metabolism of *clostridium acetobutylicum. Journal of general microbiology,* 68, 307-318.

Palazzi, E., Fabiano, B. & Perego, P. (2000). Process development of continuous hydrogen production by *Enterobacter aerogenes* in a packed column reactor. *Bioprocess Engineering,* 22, 205 – 213.

Pan, C., Yao, T., Zhaoa, P., Houa, H., (2008). Fermentative hydrogen production by the newly isolated *Clostridium beijerinckii* Fanp3. *International journal of hydrogen energy,* 33, 5383–5391.

Pan, C.M., Fan, Y.T., Xing, Y., Hou, H.W. & Zhang, M.L. (2008). Statistical optimization of process parameters on biohydrogen production from glucose by *Clostridium sp.* Fanp2. *Bioresource Technology,* 99, 3146–3154.

Park, S. H., Kim, H. J. and Cho, J. (2008). Optimal *Central Composite Designs for ftting second order response surface linear regression models. Recent Advances in Linear Models and Related Areas.* New York: Springer.

Pinchukova, E.E., Varfolomeev, S.D. & Kondrateva, E.N. (1979). Isolation, purification and study of the stability of the soluble hydrogenase from *Alcaligenes eutrophus* Z-1. *Biokhimiya, 44,* 605-615.

Rachman, M.A., Furutani, Y., Nakashimada, Y., Kakizono, T. & Nishio, N. (1997). Enhanced hydrogen production in altered mixed acid fermentation of glucose by Enterobacter aerogenes. *Journal of Fermentation Bioengineering* 83, 358-363.

Rachman, M.A., Nakashimada, Y., Kakizono, T. & Nishio, N. (1998). Hydrogen production with high yield and high evolution rate by self-flocculated cells of *Enterobacter aerogenes* in a packed- bed reactor. *Applied Microbiology*

Biotechnology, 49, 450 – 454.

Rajeshwar, K., Ibanez, J.G. and Swain, G.M. (1994). Electrochemistry and the environment. *J. Appl. Electrochem, 24*(11), 1077–1091, 1994.

Reith, J.H., Wijffels, R.H. & Barten, H. (2001). *Bio-methane & Bio-hydrogen: Status and perspectives of biological methane and hydrogen production.* Netherland: Novem.

Rhyder, R. F. (1997). *Manufacturing process design and optimization.* New York: Marcel Dekker, Inc.

Roy, R. (1990). *A primer on the taguchi method.* Competitive manufacturing series. New York: Van Nostrand Reinhold.

Safavi. S.A., Shah, F.A., Pakdel, A.K., Reza Rasoulian, G., Bandani, A.R., Butt, T.M. (2007). Effect of nutrition on growth and virulence of the entomopathogenic fungus Beauveria bassiana. *FEMS Microbiology Letters,* 270, 116–123.

Schroder, C., Selig, M. & Schonheit, P. (1994). Glucose fermentation to acetate, CO_2 and H_2 in the anaerobic hyperthermophilic eubacterium *Thermotoga maritima*: involvement of the Embden-Meyerhof pathway. *Archives of Microbiology,* 161, 460-470.

Shell International limited. (2001). *Energy Needs, Choices and Possibilities. Scenarios to 2050.* London: Shell international.

Silva, E.M., Yang, S.T., (1995). Kinetics and stability of a fibrous-bed bioreactor for continuous production of lactic from unsupplemented acid whey. *Journal of Biotechnology.* 41, 59–70.

Sim, J.H. & Kamaruddin, A.H. (2008). Optimization of acetic acid production from synthesis gas by chemolithotrophic bacterium *Clostridium aceticum* using statistical approach. *Bioresour Technol, 99,* 27–35.

Sim, J.H., Kamaruddin, A., Long, W. S. & Najafpour, G. (2007). *Clostridium aceticum*—A potential organism in catalyzing carbon monoxide to acetic acid: Application of response surface methodology. *Enzyme and Microbial Technology, 40,* 1234–1243.

Simunek, J., Kopecny, J., Hodrova, B., Bartonova, H., 2002. Identification and characterisation of *Clostridium paraputrificum*, a chitinolytic bacterium of human digestive tract. *Folia Microbiology, 47* (5), 559-564.

Simunek, J., Tishchenko, G. & Koppová, I. (2008). Chitinolytic activities of Clostridium sp. JM2 isolated from stool of human administered per orally by chitosan. *Folia Microbiololology (Praha),53*(3), 249-254.

Sinji, K., Tomoyuki, N., Yoshitaka, N., Yoshimi, B., Tai, U., Kazuo, K., Michio, K and Youichi, N. (1998). Effect of oxygen on the growth of *Clostridium butyricum* (Type Species of the Genus *Clostridium),* and the distribution of enzymes for oxygen and for active oxygen species in Clostridia. *Journal of Fermentation and Bioengineering. 86*(4), 368-372.

Sittijunda, S. & Reungsang, A. (2012). Media optimization for biohydrogen production from waste glycerol by anaerobic thermophilic mixed cultures. *International Journal of Hydrogen Energy, 37* (20), 15473–15482.

Sparling, R., Risbey, D. & Poggi-Varaldo, H.M. (1997). Hydrogen production from inhibited anaerobic composters. *International Journal of Hydrogen Energy, 22,* 563-566.

Suzuki, Y., (1982). On hydrogen as fuel gas. *International Journal of Hydrogen Energy,* 7(30),227.

Tae-Young, J., Gi-Cheol, C., Sung, H., and Suk, S. (2008). Comparison of hydrogen production by four representative hydrogen-producing bacteria. *Journal of Industrial and Engineering Chemistry, 14*(3), 333-337.

Taguchi, F., Chang, J.D., Mizukami, N., Saito-Taki, T., Hasegawa, K. & Morimoto, M. (1993). Isolation of a hydrogen producing bacterium, *Clostridium beijerinckii* strain AM 21B from termites. *Canadian Journal of Microbiology, 39*, 726-730.

Taguchi, F., Chang, J.D., Takiguchi, S. & Morimoto, M. (1992). Efficient hydrogen production from starch by a bacterium. *Journal of Fermentation and Bioengineering, 73*(3), 244–245

Taguchi, F., Hasegawa, K., Saito-Taki, T. & Hara, K. (1996). Simultaneous production of xylanase and hydrogen using xylan in batch culture of *Clostridium sp.* strain X53. *Journal of Fermentation Bioengineering, 81*, 178-180.

Taguchi, F., Mizukami, N., Saito-Taki, T. & Hasegawa, K. (1995). Hydrogen production from continuous fermentation of xylose during growth of *Clostridium sp.* strain no. 2. *Canadian Journal of Microbiology, 41*, 536 – 540.

Taguchi, F., Mizukami, N., Yamada, K., Hasegawa, K. & Saito-Taki, T. (1995). Direct conversion of cellulosic materials to hydrogen by *Clostridium sp.* strain no. 2. *Enzyme Microbial Technology, 17*, 147 – 150.

Taguchi, F., Mizukami. N., Hasegawa, K. & Saito-Taki, T. (1994). Microbial conversion of arabinose and xylose to hydrogen by a newly isolated *Clostridium sp.* No.2. *Canadian Journal of Microbiology,* 40, 228-233.

Taguchi, F., Yamada, K., Hasegawa, K., Taki-Saito, T., Hara, K. (1996). Continuous hydrogen production by *Clostridium sp.* strain No. 2 from cellulose hydrolysate in an aqueous two- phase system. *Journal of Fermentation Bioengineering, 82*, 80-83.

Tang, I.C., Okos, M.R., Yang, S.T. (1989). Effects of pH and acetic acid on homoacetic fermentation of lactate by *Clostridium formicoaceticum. Biotechnology and Bioengineering*, 34, 1063–1074.

Tanisho, S. & Ishiwata, Y. (1994). Continuous hydrogen production from molasses by the bacterium *Enterobacter aerogenes. International Journal of Hydrogen Energy, 19,* 807-812

Tanisho, S., Kuromoto, M. & Kadokura, N. (1998). Effect of CO_2 removal on hydrogen production by fermentation. *Internationl Journal of Hydrogen Energy. 23,* 559 – 563.

Tanisho, S., Suzuki, Y. & Wakao, N. (1987). Fermentative hydrogen evolution by *Enterobacter aerogenes* strain E.82005. *International Journal Hydrogen Energy, 12,* 623-627.

Tanisho, S., Wakao, S. & Kosako, Y. (1983). Biological hydrogen production by Enterobacter aerogenes. *Journal of Chemical Engineering, 16,* 529-530.

Thauer, R., Jungermann, K. and Decker, K. (1977). Energy conversion in chemotrophic anaerobic bacteria. *Bacteriol Rev, 41,* 100.

Tuhela, L., Start, A.S. & Olli, H.T. (1993). Microbial analysis of iron-related biofouling in water wells and flow cell apparatus for field and laboratory investigations. *Ground water,* 31(6), 982-988.

Turcot, J., Bisaillon, A., Hallenbeck, P. (2008). Hydrogen production by continuous cultures of *Escherchia coli* under different nutrient regimes.

International Journal Hydrogen Energy, 33, 1465–1470.

Ueno, Y., Kawai, T., Sato, S., Otsuka, S. & Morimoto, M. (1995). Biological production of hydrogen from cellulose by natural anaerobic microflora. *Journal of Fermentation Bioengineering, 79,* 395-397.

Van Ginkel, S. & Logan, B.E. (2005). Inhibition of biohydrogen production by undissociated acetic and butyric acids. *Environmental Science Technology,* 39:9351–9356.

van Niel, E.W., Budde, M.A., de Haas, G. van der W.F., Claassen, P.A. & Stams, A.J. (2002). Distinctive properties of high hydrogen producing extreme thermophiles, *Caldicellulosiruptor saccharolyticus* and *Thermotoga elfii. International Journal Hydrogen Energy, 27,* 1391-1398.

Van, S.G., Sung, S. & Lay, J.J. (2001). Biohydrogen production as a function of pH and substrate concentration. *Environ. Sci. Technology, 35,* 4726–4730.

Vandak, D., Zigová, J., Šturd'ık, E., Schlosser, S. (1997). Evaluation of solvent and pH for extractive fermentation of butyric acid. *Process Biochemistry, 32,* 245–251.

Vavilin, V.A., Rytow, S.V., and Lokshina, L.Y. (1995). Modelling hydrogen partial pressure change as a result of competition between the butyric and propionic groups of acidogenic bacteria. *Bioresource Technol, 54*(2), 171– 177.

Venkataramana, G., Anoop, S., Nagamany, N., Maung, T., (2008). Photofermentation of malate for biohydrogen production; A modeling approach. *International Journal of Hydrogen energy, 33,* 2138 – 2146.

Wang, J. & Wan, W. (2008). Effect of Fe^{2+} concentration on fermentative hydrogen production by mixed cultures. *International Journal of Hydrogen*

Energy, 33,1215–1220.

Wang, J. & Wan, W. (2009). Experimental design methods for fermentative hydrogen production: *A review. International journal of hydrogen energy,* 34, 235–244.

Wang, J.L. & Wan, W. (2008). Comparison of different pretreatment methods for enriching hydrogen-producing cultures from digested sludge. *International Journal of Hydrogen Energy, 33,* 2934–2941.

Watson, R. T., Zinyowera, M. C. & Moss, R. H. (1998). *IPCC Special Report on The Regional Impacts of Climate Change: An Assessment of Vulnerability.* Cambridge: Cambridge University Press.

Weinberg, E. D. (1989). In Regulation of secondary metabolism in Actinomycetes. CRC press: Boca Raton.

Wen-Ming, C., Ze-Jing, T., Kuo-Shing, L., Jo-Shu, C. (2005). Fermentative hydrogen production with *Clostridium butyricum* CGS5 isolated from anaerobic sewage sludge. *International Journal of Hydrogen Energy, 30,* 1063 – 1070.

Weuster-Botz, D. (2000). Experimental design for fermentation media development: statistical design or global random search? Journal of Bioscience Bioengineering. 90, 473–483.

Yang, H. & Shen, J. (2006). Effect of ferrous iron concentration on anaerobic bio-hydrogen production from soluble starch. *International Journal Hydrogen Energy,* 31:2137-2146.

Yokoi H, Tokushige T, Hirose J, Hayashi S, Takasaki Y (1998) H2 production from starch by a mixed culture of *Clostridium butyricum* and *Enterobacter aerogenes. Biotechnology Letters, 20,* 143 – 147.

Yokoi, H., Ohkawara, T., Hirose, J., Hayashi, S. & Takasaki, Y. (1995).

Characteristics of hydrogen production by aciduric *Enterobacter aerogenes* strain HO-39. *Journal of Fermentation Bioengineering, 80,* 571 – 574.

Yokoi, H., Ohkawara, T., Hirose, J., Hayashi, S., Takasaki, Y. (1995). Characteristics of hydrogen production by aciduric *Enterobacter aerogenes* strain HO-39. *Journal of Fermentation and Bioengineering,* 80, 571–574.

Yu, H. and Fang, H. (2001). Acidification of mid- and high- strength dairy wastewaters. *Water Research*, 35, 3697–3705.

Yu, J. and Takahashi, P. (2007). Biophotolysis-based Hydrogen Production by Cyanobacteria and Green Microalgae. *Communicating Current Research and Educational Topics and Trends in Applied Microbiology.* 70-88.

Yu, J., Tong, M., Sun, X., Li, B., (2007). Cysteine-modified biomass for Cd (II) and Pb (II) biosorption. *Journal of Hazardous Materials, 143,* (1–2), 277–284.

Yuan, Z., Yang, H., Zhi, X. & Shen, J. (2008) Enhancement effect of L-cysteine on dark fermentative hydrogen production. *International journal of hydrogen energy, 33,* 6535–6540.

Zhang, T., Liu, H., Fang, H. (2003). Bio-hydrogen production from starch in wastewater under thermophilic conditions. *Journal of Environmental Management, 69,* 149–156.

Zhao, X., Defeng, X., Bingfeng, L., Lu, L., Jun, Z. & Nanqi, R. (2012). The effects of metal ions and L-cysteine on hydA gene expression and hydrogen production by *Clostridium beijerinckii* RZF-1108. *International Journal of Hydrogen Energy, 37*(18), 13711-13717.

Zhao, X., Xing, D., Na F., Liu, B. & Ren, N. (2011). Hydrogen production by the newly isolated *Clostridium beijerinckii* RZF-1108. *Bioresource Technology*, 102, 8432–8436.

Zhu, Y. & Yang, S. (2004). Effect of pH on metabolic pathway shift in fermentation of xylose by *Clostridium tyrobutyricum*. *Journal of Biotechnology*, 110, 143–157.

Appendices

Appendix 1

Table 1.1 Abbreviations

Acetyl-CoA	Acetylcoenzyme A
ADP	Adenosine diphosphate
ATP	Adenosine triphosphate
CCD	Central composite design
COD	Chemical oxygen demand
CSTR	Continuous-flow stirred-tank reactor
GlcNAc	N-acetylglucosamine
HPLC	High Performance liquid chromatography
IPCC	Intergovernmental Panel on Climate Change
NADP	Nicotinamide adenine dinucleotide phosphate
PNS	Purple non-sulphur bacteria.
ppmv	Parts per million volumes
PY	Peptone yeast
RPM	Revolutions per minute

Table 1.2 Chemical formula, structure and molecular weight of N-acetylglucosamine, Glucose and organic metabolites produced.

Chemical	Formula	Chemical structure	Molecular weight (g/mol)
N-acetylglucosamine	$C_8H_{15}NO_6$		221.21
Glucose	$C_6H_{12}O_6$		180.16
Acetate	$C_2H_3O_2$		59.04
Lactate	$C_3H_5O_3$		89.07
Formate	CHO_2		45.02
Ethanol	C_2H_6O		23.08
Butyrate	$C_4H_7O_2$		87.10
Propionate	$C_3H_5O_2$		73.07

Appendix 1.1 Bacteria storage condition in broth.

Bacto thioglycolate medium without dextrose (Difco 143763) was used to store *C. beijerinckii* and *C. paraputrificum* anaerobically. Steps for preparation are as follows:

- 24 grams of the thioglycolate, was suspended in 1 liter distilled water or deionized water and allowed to mix for a minimum of 15 minutes.

- The thioglycolate solution was sterilized at 121°C for 15 minutes.

- It was stored at 15°C to 30°C.

- If stored for a long time it was reheat once in a water bath to remove the absorbed dissolved oxygen prior to use.

The formulation for the thioglycolate medium are shown in the table below:

Table 1.3 Thioglycolate medium formulations.

Formulation Constituents	Concentration (g/l)
Bacto yeast extract	5
Bacto yeast extract	15
L-cystine	0.25
Sodium Chloride	2.5
Thioglycolate acid	0.3 ml
Bacto Agar	0.75
Methylene blue	0.002

g/l = grams per litre.
ml = millilitres.

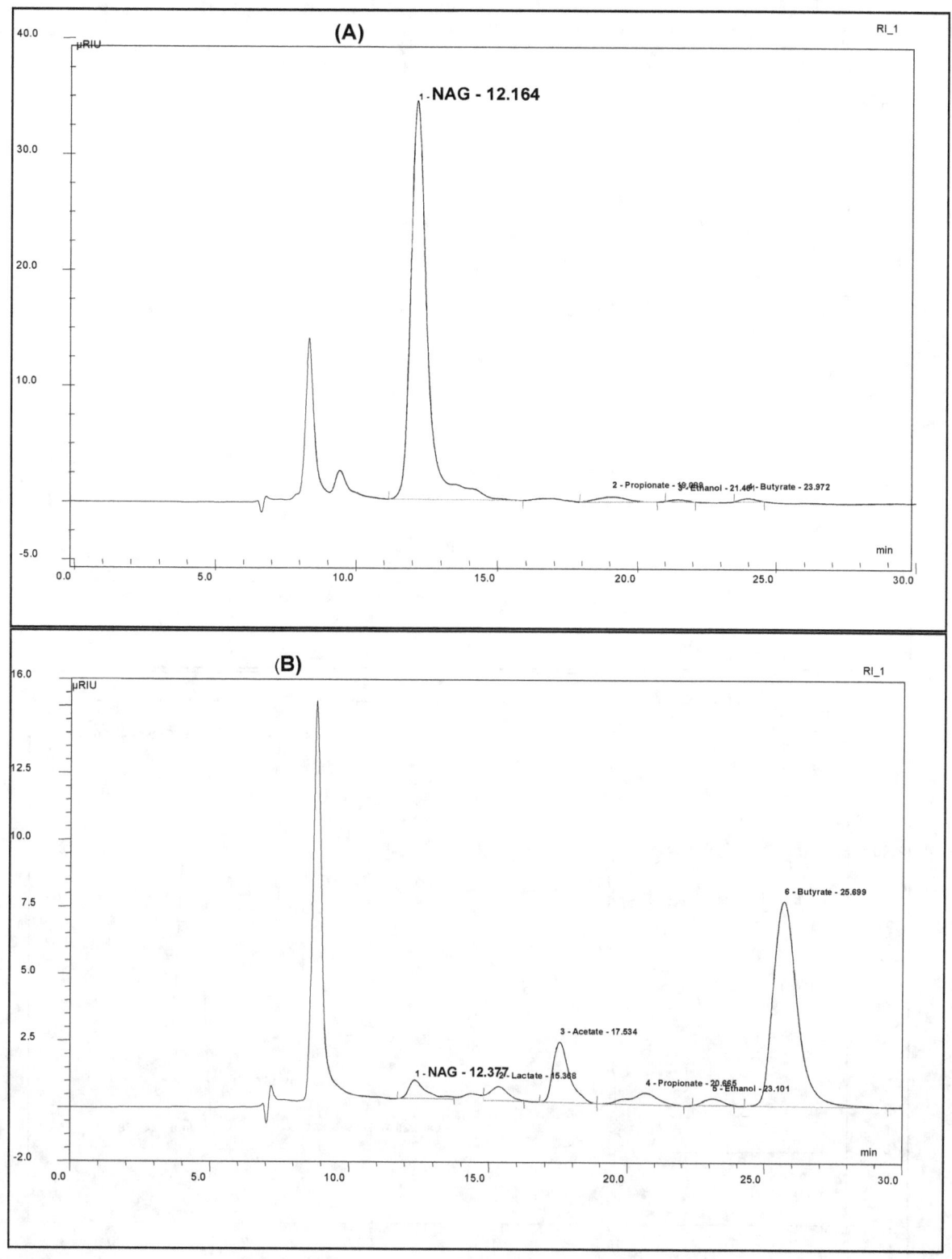

Figure 1.1 The metabolite profiles, detected by HPLC, for *C. beijerinckii* when supplied with *N*-acetyl glucosamine after: a) Day 1 incubation: b) 15 days of incubation.

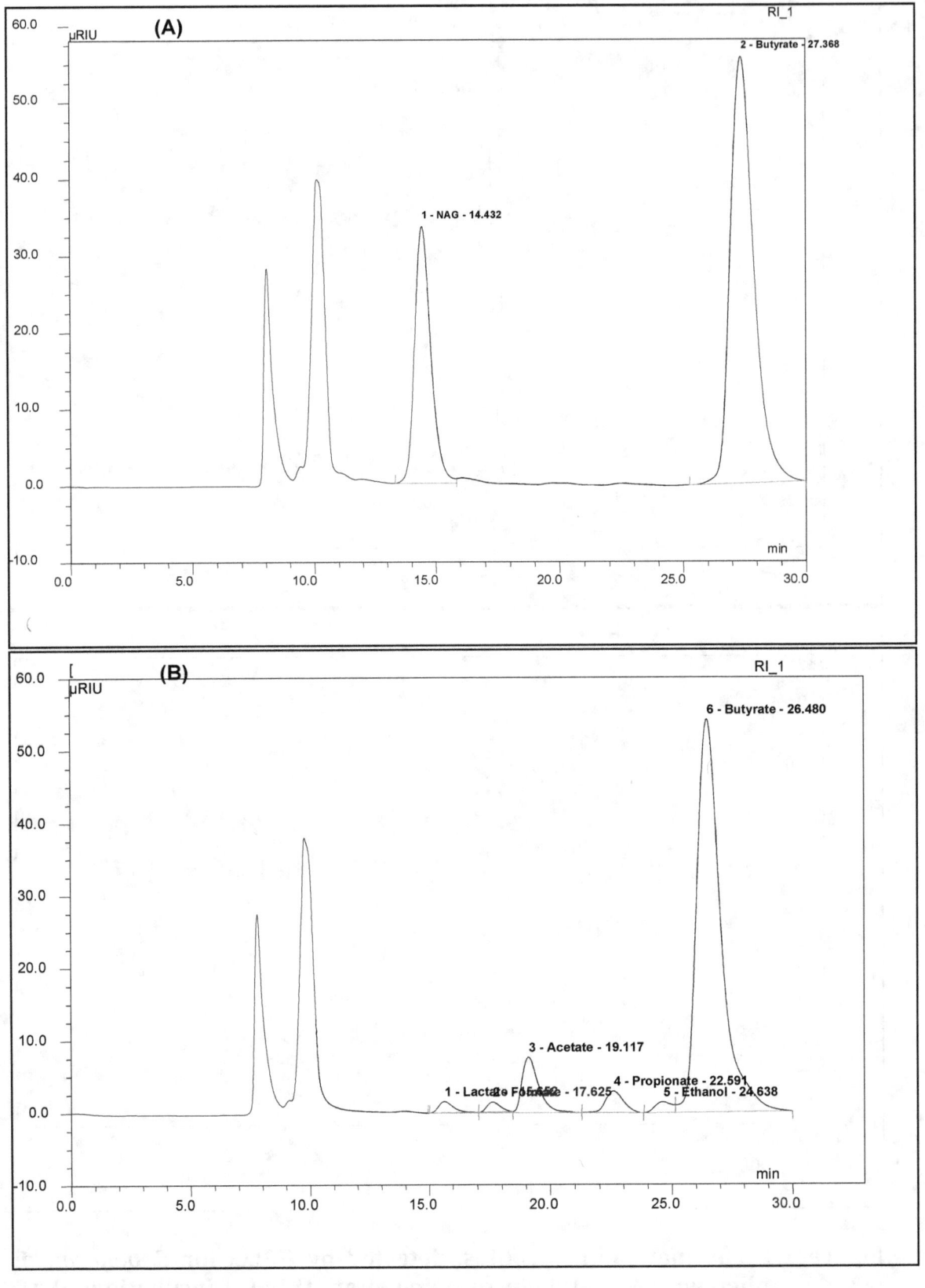

Figure 1.2 The metabolite profiles, detected by HPLC, for *C. paraputrificum* when supplied with *N*-acetyl glucosamine after: a) Day 1 incubation: b) 15 days of incubation.

Figure 1.3. Example of a syringe culture used to culture *C. beijerinckii* and *C. paraputrificum* using the statistical optimization designs.

Appendix 2. Equipment Calibration and measurement

Appendix 2.1 Culture Optical Density measurements.

Samples were collected from the cultures immediately after inoculation and after each fermentation period. The density of bacterial culture was determined by measuring the absorbance at 560 nm using the appropriate uninoculated Clostridia medium as a blank. Steps for checking the culture optical density measurements are stated below:

- The spectrometer was turned on and then allowed to proceed with the internal calibration. Enough time was given to allow the bulb to warm up before using it.

- When the instrument was ready, its absorbance wavelength was set to 560nm.

- Gloves were worn when handling the cuvette and touching the clear surface through which the laser beams will pass through to measure the absorbance. Checks were made to ensure that the clear sides were clean with no visible smear or residue. The cuvettes were cleaned using tissue when necessary.

- Blank samples were made using the appropriate uninoculated *Clostridium* medium. An appropriate quantity of water was poured into the cuvette, so that when in the spectrophotometer the liquid level is above the line of the lase beam.

- The cuvette was then carefully placed into the spectrophotometer. The side to face the laser beam was indicated by the presence of an arrow at the top of he side of the cuvette.

- The blank button (Zero) was pressed and allowed until the reading was 0.00, and then the blank was removed.

- The sample to be measured was prepared by mixing well to ensure there has been no settling of cells and that there is a homogenous distribution of cells throughout the sample vessel. An appropriate quantity of sample 1 to 1.5 ml was transferred into a clean cuvette.

- The cuvette containing the sample was immediately placed into the spectrophotometer and the reading was recorded.

- Optical density readings above 0.7 were diluted with the appropriate uninoculated clostridium medium. These readings were done in triplicate by preparing separate samples for each reading.

Appendix 2.2 Calibration of the bioreactor temperature probe

It is important to note that the temperature probe needs to be calibrated before the pH probe is calibrated; this is done because the sensitivity of the pH probe is influenced by temperature. Calibration of the temperature probe is a single point calibration done at ambient temperature. The steps for calibration are as follows:

- After connection of the temperature probe to the bioreactor control system, the bioreactor panel was put on.

- The bioreactor temperature probe and a standard pre-calibrated temperature probe were placed in a 50 ml beaker containing 30 ml of distilled water and the temperature readings displayed for the two probes were allowed to stabilize.

273

- If the temperature readings differ, the bioreactor process set point temperature was adjusted to compensate the temperature value of the calibrated thermometer.

Appendix 2.3 Bench pH Meter calibration

The bench pH meter was calibrated every day prior before use. The steps involved in the calibration process are stated bellow:

- Prior to use the temperature of the pH buffers were brought down to room temperatures. Temperatures of solutions to be used in the calibration process were within ±3 °C of room temperature.

- Calibration instructions for the bench pH meter (Mettler Toledo) were followed with University of Hertfordshire laboratory procedures.

- To begin the calibration, the pH electrode was removed from the storage solution and rinsed with deionized water. Two beakers were used for each buffer solution to avoid cross contamination during calibration. The first beaker was used to rinse the pH electrode while the other one was used for the actual pH calibration process. By applying to this technique the risk of contaminating the buffer solution is reduced and the calibration accuracy is not affected.

- The Calibrate key was pressed to initiate the calibration process, the pH meter displayed CAL.1 on the display screen.

- The electrode was placed in the beaker containing the pH 4.00 buffer and then stirred at a slow speed. The pH meter is ready for the next step of calibration when the meter signifies that a stable reading has been achieved for the pH 4.00 buffer (pH icon appears). The pH electrode was removed from the buffer and then rinsed with deionized water and blot dry, then the calibrate key was pressed again and then CAL.2 was displayed on the screen. At this point the pH meter required the pH 7 or 10 buffer to end the calibration process.

- The step above was repeated for the pH 7.00 buffer. After the pH 7 buffer has been used for the calibration process and the pH icon displayed again. The pH meter will display the electrode slope value at the end of the calibration process. Note that on any occasion where the slope value was less than 95% the calibration process was repeated again.

Appendix 2.4 pH Measurements

After calibration of bench pH meter, the next step for pH measurement are stated below:

- The calibrated probe was removed from 3M KCl storage solution and then rinsed with distilled water then blot dry.
- The pH probe is then placed in the solution to be measured, then the pH reading is allowed to stabilize and the pH value recorded.
- The pH probe is rinse again and blot dry. The probe is then placed in the next solution to be measured or processed to shut down
- After the last pH measurement, the pH probe is rinsed and blot dry then put in a 3M KCl solution for storage.

Appendix 2.5 Calibration of the Bioreactor pH – Electrode

The bioreactor pH-electrode is calibrated using a two-point calibration, which determines the electrode parameters zero drift and slope using buffer solutions. The calibration of the bioreactor pH electrode is always done before it is attached to the culture vessel. The same technical buffers used for the calibration of the bench pH meter were used for the calibration of the bioreactor pH. Protective gloves were worn when the pH buffers were handled during the calibration procedure.

The software of the control system in the bioreactor control panel calculates the PH-value derived from the electrode voltage based on the zero drift and slope, in relation to the Nernst equation. The actual temperature of the calibrating condition can be entered manually during calibration to compensate for the effects of temperature.

Prior to sterilization the pH electrode is calibrated before attaching it to the fermentation vessel. During the sterilization process the high temperature involved in the process can cause a shift in the electrode zero. Furthermore chemical components in the culture medium can affect the ability of the electrode to measure pH values accurately. For this reason the pH electrode was calibrated properly before use and also checked regularly to confirm its accuracy.

However the software application used on the control system provides a recalibration function, which helps to compensate for changes in measurement values of the pH-electrode. By using this recalibration key pH values can be measured offline in a sample from the fermentation process and then the value entered into the calibration menu. The control system then recalculates the calibration function and then applies the new adjustment to the pH control function. This process was carried out every time a fermentation experiment was carried out on the vessel.

Fresh samples were always used in the recalibration process and care was taken to avoid any influence on the sample that may make the recalibration process inaccurate. The pH of the fresh samples were measured quickly so it can be compared with the bioreactor pH, this was done quickly to avoid drifts in pH during the time lapsed during sampling as culturing pH may change with time. Offline bench pH meter was always validated to ensure accuracy.

Appendix 2.6 pH-Calibration Routine

- To initiate pH calibration, the main function key calibration was pressed to select the menu for the control process. In the control menu the cursor displays on the manual or automatic entry.

- ALTER was pressed to select manual temperature compensation and then it was confirmed with ENTER. The cursor automatically jumped to the TEMP-entry, then the reference temperature labeled on the buffer bottle was inserted on the panel for the relationship of pH versus temperature controls.

- The pH electrode was placed in a beaker containing the zero buffer of pH 7.0. The temperature of the buffer was measured and the reference temperature was entered into the temperature entry slot and then confirmed by pressing the enter button, then the cursor jumped to the BUFZ entry.

- The pH value (the reference pH-value of the zero buffer labelled on the bottle for the actual buffer temperature) was entered in the BUFZ-ENTRY and then confirmed with the enter key. The control system then starts an automatic monitoring of the electrode when the cursor jumps to ok. If the measured signal is stable within the given limits, the system stores the measured voltage as the electrode's zero. Then the cursor automatically jumps to the BUFS entry.

- The electrode was then removed from the beaker containing the zero buffer (buffer at pH 7) and then it was rinsed with deionized water and blotted dry then it was placed in the slope buffer.

- The pH value (the reference pH-value of the slope buffer labeled on the bottle for the actual buffer temperature) was entered in the BUFS-ENTRY and then confirmed with the enter key. The control system then starts an automatic monitoring of the pH electrode when the cursor jumps to ok. If the measured signal is stable within the given limits, the system stores the measured voltage as the electrode's slope.

- At this point when the system stabilizes, the cursor automatically jumps to the manual /automatic entry and switches over the temperature. After this happens the system has been calibrated. This calibration processes was done every time before a fermentation process was carried out.

Appendix 2.7 Calibration of the dissolved oxygen (pO₂) - electrode

The calibration of the dissolve oxygen electrode is done in percentage oxygen saturation, % pO₂. The calibration is usually done in a two-point calibration. Generally a standard solution without oxygen provides the zero current and a solution saturated with oxygen is known as 100% saturated. The pO2 measurements are calculated using the resulting electrode parameters zero current and slope.

The dissolved oxygen electrode is calibrated after it has been attached to the culture vessel and also after sterilization. To zero the dissolve oxygen electrode during calibration, the culture medium was gassed with nitrogen to displace the dissolved oxygen in the medium. To saturate the medium with oxygen to give an electrode slope of 100%, the medium was gassed with air.

The actual dissolved oxygen saturation is displayed as percentage saturation % pO2 on the calibration menu for the dissolved oxygen pO₂ –electrode.

Appendix 2.8 Steps of operation for calibrating dissolved oxygen electrode.

- At the initiation of the dissolved oxygen calibration, the vessel medium is not purged with any gas containing oxygen. To zero the vessel during the calibration process using the supply of oxygen free gas, the medium was gassed with nitrogen until the measured dissolved oxygen concentration is stable and equal or close to 0% dissolved oxygen concentration.

- The temperature of the culture vessel is also inserted into the control system for accuracy.

- The measured value for the nitrogen gassing out is confirmed by pressing the enter key on the control system. At this point the system starts an automatic monitoring of the dissolved oxygen electrode. Once the measured electrode's zero current is stable, the system stores this as zero reference. The cursor then moves to the next step of calibration, it automatically moves to the AIR entry.

- The culture vessel is then gassed with air at a flow rate of 1l/min. The saturation point of 100% was confirmed by pressing the enter key on the air entry. At a point when the slope reference is not 100% the actual value was entered into the system.

- At this step the cursor then jumps to ok and the system starts an automatic monitoring of the electrode. If the measured signal is stable within the given limits, the system stores this as slope. At this point the system dissolved oxygen electrode is calibrated.

- During the process the system calculates the dissolved oxygen pO_2 –value from the measured electrode's current in relation to the calibrated zero, slope and the temperature.

Appendix 2.9 Preparing the culture vessels and accessories

Appendix 2.9.1 Checking the equipment

During assembly of the culture vessel and mounting of the probes and accessories, all parts were carefully checked for damages. The glass vessel was examined thoroughly; the seals on the top plate and the O-rings of all mountings were checked.

Damaged seals can be the cause of contaminations of the culture fermentation process. Damages seals on accessories were replaced often before the fermentation run.

Appendix 2.9.2 Assembling and equipping the culture vessel

- The top-plate of the culture vessel was unscrewed and the top plate dismounted. The o-ring was cleaned and when damaged was replaced.

- Parts like the sparger line were installed to the top plate from below and the ports with no need for attachments of accessories were fitted with blind closures.

- All seals were checked and damaged ones were replaced. The seals were also greased to prevent gluing on contact surfaces.

- The top plate was then mounted back on the glass vessel and then the screws were hand tightened. Care was taken to avoid uneven pressure applied to the rim of the glass vessel when tightened.

- All probes and accessories for the fermentation run were then attached to the top plate. Prior to attaching the pH electrode to the top plate, it was calibrated, but the calibration of the dissolved oxygen pO_2 probe was carried out carried out in oxygen-free culture medium after autoclaving.

- A short piece of tubing was attached to the membrane filter and attached to the air inlet and the other part of the filter attached to the sparger pipe in the top plate.

- The condenser was then screwed directly to the plate and then a membrane filter attached to the exhaust. This was done with the aid of another silicon tube.

- Aluminum foil was wrapped around the exposed electrical connectors and all tubing of the parts (for example inlet air tubing, harvest pipe), which reached the culture solution, were clamped.

- Prior to autoclaving the exhaust line was kept open so that pressure compensation with ambient pressure can take place, as it is necessary during and after the sterilization in the autoclave.

Appendix 2.9.3 Filling the Jacket of the culture vessel

It is important to note that the thermostat liquid in the jacket of the bioreactor culture vessel provides optimum heat transfer during the process of sterilization. Hence, prior to sterilization of the culture vessel or before any fermentation run, the culture vessel jacket was checked and refilled when necessary with the thermostat liquid. This process was carried out as stated in the steps below:

- The thermostat tubing was connected to the culture vessel, by attaching the outlet supply unit to the lower quick connector at the culture vessel jacket inlet. The upper connector at the culture vessel outlet was attached to the inlet supply unit.

- The power main of the basic unit was switched on and the fill thermostat switch was pressed to fill the thermostat system, tank and the jacket of the culture vessel. After the jacket was filled with the thermostat liquid, it was disconnected again to keep the vessel for sterilization. The upper outlet plug on the jacket vessel was not closed during sterilization as the thermostat liquid expands during the heating up process in the autoclave. The excess liquid in the jacket was allowed to flow out of the jacket as overpressure build up in the jacket may damage the vessel.

Appendix 2.9.4 Filling the culture vessel

- Culture solution was poured through any of the opening ports on the top plate of the vessel.

- Other medium components that were not autoclavable for example N-acetyl glucosamine and Iron sulphate were filter sterilized into the medium after autoclaving.

- The inoculum used in the fermentation run was poured through any of the opening ports on the top plate of the vessel.

Appendix 2.9.5 Sterilization of the culture vessel

- Care was taken to ensure that all open ports in the top plate of the culture vessel are closed. All loose parts were kept or attached to the fermenter vessel with autoclave tapes so they don't get lost during handling and transportation of the vessel

- All the electrode plugs were covered with aluminum foil to protect them from the impact of steam. The pH and dissolved oxygen electrodes all have special protective caps, this were used.

- The jacket of the culture vessel was completely filled with thermostat liquid (water).

- The culture vessel and all its working accessories were autoclaved at a temperature of 121°C for 15 minutes.

Appendix 2.9.6 Preparing a fermentation run

- After sterilization the culture vessel and the connected equipment in the autoclave were allowed to cool down. Before removing the culture vessel from the autoclave, protective gloves were worn when necessary.

- The culture vessel was transported carefully to the work place for further process set up. Care was taken during transport to prevent loosening of tubes or assemblies connected. However tubes were cable tied to the ports on the plate.

- Agitator motor was mounted onto the stirrer shaft coupling at the top plate of the culture vessel. The gas/air supply was connected to the inlet gas/air filter.

- The thermostat system was connected to the culture vessel. The cooling water supply and back flow tubing were connected to the exhaust cooler.

- Protective aluminum foils were removed from the electrode connectors and the specific cables for the electrodes and corresponding sockets were connected to the control unit.

- The system was switched on and the thermostat liquid in the jacket of the culture vessel was topped up

- The operation parameters at the measurement and control system were checked and confirmed.

- The culture media was then inoculated with the starter culture. Prior to inoculation the dissolved oxygen calibration was carried out as stated in appendix 2.8 above and the culturing temperature (37°C) was allowed to stabilize before inoculation was done

Appendix 2.9.7 Connecting air/ nitrogen supplies

It is important to note that the dissolved oxygen pO_2-electrode must be calibrated before starting the nitrogen supply. The calibration process is explained in details in appendix 2.8.

- The air supply was removed form the air supply tubing from its holder on the vessel, and then attached to the filter membrane. Any gas supplied through the filter membrane into the culture vessel is filter sterilsed.

Appendix 2.9.8 Connecting the Exhaust Cooler

- The cooling water supply and drain were connected to the exhaust cooler. There is usually a constant flow of cooling water through the exhaust cooler through out the fermentation run.

Appendix 2.9.10 Connecting Probes and Setting the Measurement and Control System

Overview on Connection and Pre-adjustment

- The electrode cables were connected to the electrodes and then to the corresponding input plugs on the control unit.
- The control unit was then switched on and then the equipment was checked to ensure its proper functioning.
- The required functions were selected and the necessary parameters were adjusted on the control system. Agitation was set at 150 rpm and temperature was set at 37°C.

Appendix 2.10.0 Carrying-out a fermentation run

Appendix 2.10.1 Sterility Test

Damaged bioreactor culture vessel parts, particularly faulty seals, incorrect process run handling while equipping the culture vessels before and after the sterilisation process, can cause contamination of the culture medium. This usually causes fermentation runs to fail and leave results unachieved. To detect signs of incorrect handling or faults, especially regarding seals a sterility test was done each time before a fermentation run was carried out. Sterility test helps to prevent unknown time wastage in a fermentation run. Steps for the sterility test are explained below:

- Simply mimicking the real fermentation run without inoculation carries out the test. All necessary components were connected to the culture vessel and the necessary operating conditions for the fermentation run adhered too.

- The process was allowed to run for a 24 h test period. This allowed duration is sufficient to detect most infections by common environmental germs. A sample was taken out from he vessel and checked under the microscope for contamination.

- If there was contamination the vessel was autoclaved again and the preparation methods for fermentation run started again. That is reassembling and preparing the culture vessel again for the planned fermentation run as infection occurred.

- If the sterility test is successful and there were no contaminants the fermentation run was allowed to continue as planned.

Appendix 2.10.2 Post autoclave checks and calibrations.

The post-autoclave checks are performed to confirm that dissolved Oxygen probe calibration was not affected during the sterilization process by autoclaving. After the culture vessel has been autoclaved and all accessories attached and all control parameters set and running, the stirrer set at 150 rpm and temperature condition at 37°C.

Nitrogen supply from the cylinder to the bioreactor culture vessel was connected through a filter membrane then into the culture medium. During the supply of nitrogen the dissolved oxygen value was monitored and checked if the value reduced slowly within 15 minutes and at a point when it was not the case it was a sign that there was a possible damage to the dissolved oxygen membrane.

When the dissolved oxygen value reduced and stabilized at a value of $0\pm 5\%$ the nitrogen tap was closed to stop the supply of nitrogen.

Airline was connected and the culture medium supplied with air through a filter vessel to also monitor the response to oxygen as the dissolved oxygen value on the bioreactor panel was monitored. The dissolved oxygen value should go up within 2 minutes. Once the dissolved oxygen value increases and its stable the air supply was turned off. At this point the post autoclave check is complete.

However, just before inoculation the dissolved oxygen value was always reduced to 0 % so as to attain an anaerobic environment for the bacteria.

Appendix 2.10.3 Pre-inoculation checks

After the culture vessel has been autoclaved and all accessories attached and all control parameters set and running, the stirrer set at 150 rpm and temperature condition at 37C. The sample area was sprayed with 70% ethanol and then allowed to dry. The sample bottle was checked if screwed on tightly and 5 to 10 ml of the sample was withdrawn from the bioreactor into the sample bottle. This process was done aseptically.

The sample bottle was closed with the lid immediately after sampling and then the pH of the sample was measured with the offline pH meter.

The online pH and the offline pH readings were compared and adjustment made to the online pH when necessary. If the pH difference was up to ±0.2 the online pH value was adjusted to the offline value.

Steps were taken to ensure that the temperature of the medium reached the set values before recalibrating the pH with the offline value and also before inoculation.

Appendix 2.10.4 Inoculating the Culture Vessel

Inoculation of the culture vessel was done by using a sterile syringe. Inoculation was 10% of the total media volume. The steps taken for inoculation are explained below:

- The required volume of solution from the inoculum culture was removed with a sterile 50ml syringe and then transferred into the culture vessel aseptically to avoid contamination.

- The blind closure from the inoculation port equipped with a septum membrane was removed. The port was flamed using a Bunsen burner to prevent contamination of the membrane. The membrane was also sprayed with IMS and then allowed to dry.

- The syringe containing the inoculum was then fit with a sterile needle and then used to puncture the membrane in order to transfer the inoculum solution into the culture vessel.

Appendix 2.10.5 Transferring Solutions by Means of a Syringe

- The sterile solution containing the medium components was removed from the storage bottle into a syringe. A sterile needle attached to a sterile filter was then connected to the syringe.

- The next step followed as explained in appendix 2.10.4 for inoculation using a syringe.

Appendix 2.11 HPLC standards

Lactate, formate, acetate, propionate, ethanol and butyrate were used in the forms of sodium lactate (Sigma, 71716), sodium formate (Sigma 17841), sodium acetate (BDH 102363P), sodium propionate (Sigma, P1880) and sodium butyrate (Sigma, B5887) respectively. Ethanol was used as 100% ethanol (Sigma-Aldrich E7023). The molecular weights of the sugars, ethanol, and the acids and their salts are shown in table 2.1

Table 2.1 Molecular weights of sugars and acids used as metabolite standards.

Sugar & Metabolites	Molecular weights	Actual molecular weight
Glucose	180.16	180.16
N-acetylglucosamine	221.21	221.21
Lactate	89.07	112.06
Formate	45.02	68.01
Acetate	59.04	82.03
Propionate	73.07	96.06
Ethanol	23.08	23.08
Butyrate	87.10	110.19

The actual molecular weights represent the sodium salts, which include sodium lactate, sodium formate, sodium acetate, sodium propionate and sodium butyrate.

A stock solution (50 ml) containing a mixture of the sugars (Glucose and GlcNAc) and the metabolic end product standards (lactate, formate, acetate, ethanol, propionate and butyrate) were made as shown in table 2.2.

Table 2.2 Concentrations of sugar and organic acid stocks in the HPLC standard mixture.

HPLC standards	Concentration of standards (g/l)		Concentration of standards in 50 ml	
	Compounds (g)	Actual sodium compounds (g)	Compound (g)	Actual sodium compound (g)
Glucose	5	5	0.25	0.25
N-acetylglucosamine	5	5	0.25	0.25
Lactate	10	12.58	0.5	0.629
Formate	10	15.1	0.5	0.755
Acetate	10	13.88	0.5	0.694
Ethanol	5	5.01	0.25	0.251
Butyrate	5	6.32	0.25	0.316
Propionate	10	13.14	0.5	0.657

g/l = grams per litre.
ml = millilitres.

The stock solutions were stored at 4°C in a refrigerator (Labcold sparkfree). Dilutions of the stock standard solution were made and analysed in triplicate on the HPLC. A standard calibration curve was then determined to permit comparison with samples from the cultures, and to determine the concentration of sugar remaining in the medium and the concentration of the selected metabolites excreted. An Aminex HPX-87H ion exchange column (300 mm by 7.8 mm Bio rad HPLC organic acid column) was used to analyse all of the samples. The HPLC standards used were analysed in triplicate.

Appendix 3

Appendix 3.1 Plackett and Burman design for *C. beijerinckii* and *C. paraputrificum*.

Table 3.1. Total volume of biogas in triplicate for *C. beijerinckii* cultures after 20-day incubation with N-acetylglucosamine as the main carbon source.

Culture Trials	Biogas volume (ml)			Biogas average (ml)	S.D	S.E
	a	b	c			
T1	11.00	12.00	9.00	10.67	1.53	0.88
T2	16.50	16.50	22.50	18.50	3.46	2.00
T3	0.00	0.00	0.00	0.00	0.00	0.00
T4	11.00	20.00	12.00	14.33	4.93	2.85
T5	14.00	14.00	10.00	12.67	2.31	1.33
T6	20.50	18.50	18.50	19.17	1.15	0.67
T7	23.00	20.50	23.50	22.33	1.61	0.93
T8	0.00	0.00	0.00	0.00	0.00	0.00
T9	31.00	31.00	32.00	31.33	0.58	0.33
T10	0.00	0.50	4.00	1.50	2.18	1.26
T11	31.00	32.00	32.50	31.83	0.76	0.44
T12	16.00	16.50	24.50	19.00	4.77	2.75

ml = millilitres.
S.D = Standard deviation.
S.E = Standard error.

Table 3.2. Total volume of biogas in triplicate for *C. paraputrificum* cultures after 20-day incubation with N-acetylglucosamine as the main carbon source.

Culture Trials	Biogas volume (ml)			Biogas Average (ml)	S.D	S.E
	a	b	c			
T1	1.00	1.00	1.00	1.00	0.00	0.00
T2	2.00	2.00	2.00	2.00	0.00	0.00
T3	2.00	11.00	9.00	10.00	1.41	1.00
T4	1.00	2.00	2.00	1.67	0.58	0.33
T5	0.00	0.00	0.00	0.00	0.00	0.00
T6	6.00	6.00	7.00	6.33	0.58	0.33
T7	1.00	4.00	2.00	3.00	1.41	1.00
T8	4.00	4.00	5.00	4.33	0.58	0.33
T9	12.00	15.00	12.00	13.00	1.73	1.00
T10	2.00	2.00	1.00	1.67	0.58	0.33
T11	0.00	0.00	0.00	0.00	0.00	0.00
T12	0.00	0.00	0.00	0.00	0.00	0.00

ml = millilitres.
S.D = Standard deviation.
S.E = Standard error.

Table 3.3. Biomass concentration in triplicate for *C. beijerinckii* cultures after 20-day incubation with N-acetylglucosamine as the main carbon source.

Culture Trials	Final O.D			O.D Average	S.D	S.E
	a	b	c			
T1	0.525	0.580	0.731	0.612	0.11	0.06
T2	0.625	0.361	0.379	0.455	0.15	0.09
T3	0.371	1.623	0.376	0.374	0.00	0.00
T4	0.590	0.572	0.598	0.587	0.01	0.01
T5	0.552	0.560	0.550	0.554	0.01	0.00
T6	0.731	0.984	1.018	0.911	0.16	0.09
T7	0.862	0.932	0.810	0.868	0.06	0.04
T8	0.402	0.401	0.420	0.408	0.01	0.01
T9	1.314	1.275	1.564	1.384	0.16	0.09
T10	0.581	0.662	0.625	0.623	0.04	0.02
T11	1.940	1.548	1.920	1.803	0.22	0.13
T12	0.752	0.703	0.705	0.720	0.03	0.02

O.D = Culture optical density in Absorbance unit.
S.D = Standard deviation.
S.E = Standard error.

Table 3.4 Biomass concentration in triplicate for *C. paraputrificum* cultures after 20-day incubation with N-acetylglucosamine as the main carbon source.

Culture Trials	Final O.D			O.D Average	S.D	S.E
	a	b	c			
T1	0.168	0.207	0.181	0.185	0.02	0.01
T2	0.381	0.481	0.302	0.388	0.09	0.05
T3	0.689	0.790	0.680	0.720	0.06	0.04
T4	0.281	0.349	0.319	0.316	0.03	0.02
T5	0.155	0.101	0.117	0.124	0.03	0.02
T6	0.566	0.572	0.614	0.584	0.03	0.02
T7	0.391	0.398	0.420	0.403	0.02	0.01
T8	0.418	0.551	0.491	0.487	0.07	0.04
T9	0.867	1.070	0.989	0.975	0.10	0.06
T10	0.318	0.338	0.278	0.311	0.03	0.02
T11	0.131	0.103	0.116	0.117	0.01	0.01
T12	0.171	0.123	0.098	0.131	0.04	0.02

O.D = Culture optical density in Absorbance unit.
S.D = Standard deviation.
S.E = Standard error.

Table 3.5 Culture final pH in triplicate for *C. beijerinckii* cultures after 20-day incubation with N-acetylglucosamine as the main carbon source.

Culture Trials	Final pH			Final pH Average	S.D	S.E
	a	b	c			
T1	5.56	6.10	6.17	5.94	0.33	0.19
T2	6.21	6.17	6.20	6.19	0.02	0.01
T3	6.96	6.15	6.90	6.67	0.45	0.26
T4	6.63	6.64	6.63	6.63	0.01	0.00
T5	6.55	6.52	6.53	6.53	0.02	0.01
T6	6.25	6.21	6.22	6.23	0.02	0.01
T7	6.42	6.40	6.39	6.40	0.02	0.01
T8	6.88	6.86	6.87	6.87	0.01	0.01
T9	5.97	5.95	5.96	5.96	0.01	0.01
T10	6.55	6.79	6.70	6.68	0.12	0.07
T11	5.98	6.03	5.97	5.99	0.03	0.02
T12	6.10	6.10	6.09	6.10	0.01	0.00

S.D = Standard deviation.
S.E = Standard error.

Table 3.6. Culture final pH in triplicate for *C. paraputrificum* cultures after 20-day incubation with N-acetylglucosamine as the main carbon source.

Culture Trials	Final pH a	b	c	Final pH Average	S.D	S.E
T1	5.80	5.75	5.68	5.74	0.06	0.03
T2	5.90	5.90	5.88	5.89	0.01	0.01
T3	6.37	6.32	6.24	6.31	0.07	0.04
T4	6.59	6.60	6.58	6.59	0.01	0.01
T5	6.44	6.45	6.45	6.45	0.01	0.00
T6	6.09	6.10	6.10	6.10	0.01	0.00
T7	6.35	6.34	6.43	6.37	0.05	0.03
T8	6.48	6.42	6.41	6.44	0.04	0.02
T9	5.64	5.53	5.55	5.57	0.06	0.03
T10	6.57	6.59	6.58	6.58	0.01	0.01
T11	6.53	6.52	6.54	6.53	0.01	0.01
T12	6.04	6.53	6.54	6.37	0.29	0.17

S.D = Standard deviation.
S.E = Standard error.

Appendix 4

Appendix 4.1 One-factor cysteine optimization culture designs.

Table 4.1 Total biogas volumes in triplicate for *C. beijerinckii* cultures with N-acetylglucosamine as the main carbon source using one factor cysteine optimization design.

Culture Trials	Biogas volume (ml)			Biogas average (ml)	S.D	S.E
	a	b	c			
C1	49.00	50.00	51.00	50.00	1.00	0.58
C2	48.00	55.00	47.00	50.00	4.36	2.52
C3	52.00	54.00	58.00	54.67	3.06	1.76
C4	48.00	49.00	48.00	48.33	0.58	0.33
C5	42.00	44.00	43.00	43.00	1.00	0.58
C6	39.00	39.00	41.00	39.67	1.15	0.67
C7	40.00	40.00	39.00	39.67	0.58	0.33
C8	32.00	38.00	37.00	35.67	3.21	1.86
C9	34.00	36.00	36.00	35.33	1.15	0.67
C10	41.00	39.00	36.00	38.67	2.52	1.45

ml = millilitres.
S.D = Standard deviation.
S.E = Standard error.

Table 4.2. Biomass concentration in triplicate for *C. beijerinckii* cultures with N-acetylglucosamine as the main carbon source using one factor cysteine optimization design.

Culture Trials	Final O.D			O.D Average	S.D	S.E
	a	b	c			
C1	2.109	1.947	1.863	1.973	0.13	0.07
C2	2.376	2.379	2.352	2.369	0.01	0.01
C3	2.460	2.036	2.652	2.383	0.32	0.18
C4	2.224	2.024	2.016	2.088	0.12	0.07
C5	1.792	1.824	1.708	1.775	0.06	0.03
C6	1.556	1.820	1.752	1.709	0.14	0.08
C7	1.860	1.616	1.584	1.687	0.15	0.09
C8	1.692	1.411	1.784	1.629	0.19	0.11
C9	1.572	1.300	1.792	1.555	0.25	0.14
C10	1.424	1.856	1.656	1.645	0.22	0.12

O.D = Culture optical density in absorbance unit.
S.D = Standard deviation.
S.E = Standard error.

Table 4.3 Culture final pH in triplicate for *C. beijerinckii* cultures with N-acetylglucosamine as the main carbon source using one factor cysteine optimization design.

Culture Trials	Final pH a	b	c	Final pH Average	S.D	S.E
C1	5.72	5.75	5.74	5.74	0.02	0.01
C2	5.68	5.67	5.67	5.67	0.01	0.00
C3	5.71	5.71	5.70	5.71	0.01	0.00
C4	5.72	5.72	5.72	5.72	0.00	0.00
C5	5.74	5.75	5.75	5.75	0.01	0.00
C6	5.85	5.79	5.77	5.80	0.04	0.02
C7	5.80	5.81	5.80	5.80	0.01	0.00
C8	5.80	5.80	5.85	5.82	0.03	0.02
C9	5.90	5.87	5.82	5.86	0.04	0.02
C10	5.87	5.79	5.83	5.83	0.04	0.02

S.D = Standard deviation.
S.E = Standard error

Table 4.4 Statistical analysis between independent variable L-cysteine.HCl.H$_2$O and dependent variables Total biogas volume, biomass, culture final pH, acetate and butyrate.

Source	Dependent Variable	Sig.
	Biogas	.000[a]
	biomass	.000[b]
Corrected Model	ph	.000[c]
	acetate	.000[d]
	butyrate	.000[e]
	Biogas	.000
	biomass	.000
Intercept	ph	.000
	acetate	.000
	butyrate	.000
	Biogas	.000
	biomass	.000
cysteine	ph	.000
	acetate	.000
	butyrate	.000
Error	Biogas	
	biomass	

	ph					
	acetate					
	butyrate					
	Biogas					
	biomass					
Total	ph					
	acetate					
	butyrate					
	Biogas					
	biomass					
Corrected Total	ph					
	acetate					
	butyrate					

a. R Squared = .925 (Adjusted R Squared = .891)

b. R Squared = .799 (Adjusted R Squared = .708)

c. R Squared = .893 (Adjusted R Squared = .844)

d. R Squared = .948 (Adjusted R Squared = .925)

e. R Squared = .956 (Adjusted R Squared = .936)

Table 4.5 Post Hoc, Tukey HSD multiple comparison statistical analysis between independent variable L-cysteine.HCl.H$_2$O and dependent variables Total biogas volume, biomass, culture final pH, acetate and butyrate.

Dependent Variable	(I) cysteine	(J) cysteine	Mean Difference (I-J)	Std. Error	Sig.	95% Confidence Interval	
						Lower Bound	Upper Bound
Biogas	0 g/l	0.5 g/l	.0000	183.18176	1.000	-648.6663	648.6663
		1 g/l	-466.6667	183.18176	.303	-1115.3330	181.9997
		1.5 g/l	166.6667	183.18176	.994	-481.9997	815.3330
		2 g/l	700.0000*	183.18176	.028	51.3337	1348.6663
		2.5 g/l	1033.3333*	183.18176	.001	384.6670	1681.9997
		3 g/l	1033.3333*	183.18176	.001	384.6670	1681.9997
		3.5 g/l	1433.3333*	183.18176	.000	784.6670	2081.9997
		4 g/l	1466.6667*	183.18176	.000	818.0003	2115.3330
		4.5 g/l	1133.3333*	183.18176	.000	484.6670	1781.9997
	0.5 g/l	0 g/l	.0000	183.18176	1.000	-648.6663	648.6663
		1 g/l	-466.6667	183.18176	.303	-1115.3330	181.9997
		1.5 g/l	166.6667	183.18176	.994	-481.9997	815.3330

298

	2 g/l	700.0000*	183.18176	.028	51.3337	1348.6663
	2.5 g/l	1033.3333*	183.18176	.001	384.6670	1681.9997
	3 g/l	1033.3333*	183.18176	.001	384.6670	1681.9997
	3.5 g/l	1433.3333*	183.18176	.000	784.6670	2081.9997
	4 g/l	1466.6667*	183.18176	.000	818.0003	2115.3330
	4.5 g/l	1133.3333*	183.18176	.000	484.6670	1781.9997
	0 g/l	466.6667	183.18176	.303	-181.9997	1115.3330
	0.5 g/l	466.6667	183.18176	.303	-181.9997	1115.3330
	1.5 g/l	633.3333	183.18176	.059	-15.3330	1281.9997
	2 g/l	1166.6667*	183.18176	.000	518.0003	1815.3330
1 g/l	2.5 g/l	1500.0000*	183.18176	.000	851.3337	2148.6663
	3 g/l	1500.0000*	183.18176	.000	851.3337	2148.6663
	3.5 g/l	1900.0000*	183.18176	.000	1251.3337	2548.6663
	4 g/l	1933.3333*	183.18176	.000	1284.6670	2581.9997
	4.5 g/l	1600.0000*	183.18176	.000	951.3337	2248.6663
	0 g/l	-166.6667	183.18176	.994	-815.3330	481.9997
	0.5 g/l	-166.6667	183.18176	.994	-815.3330	481.9997
	1 g/l	-633.3333	183.18176	.059	-1281.9997	15.3330
	2 g/l	533.3333	183.18176	.166	-115.3330	1181.9997
1.5 g/l	2.5 g/l	866.6667*	183.18176	.004	218.0003	1515.3330
	3 g/l	866.6667*	183.18176	.004	218.0003	1515.3330
	3.5 g/l	1266.6667*	183.18176	.000	618.0003	1915.3330
	4 g/l	1300.0000*	183.18176	.000	651.3337	1948.6663
	4.5 g/l	966.6667*	183.18176	.001	318.0003	1615.3330
	0 g/l	-700.0000*	183.18176	.028	-1348.6663	-51.3337
	0.5 g/l	-700.0000*	183.18176	.028	-1348.6663	-51.3337
	1 g/l	-1166.6667*	183.18176	.000	-1815.3330	-518.0003
	1.5 g/l	-533.3333	183.18176	.166	-1181.9997	115.3330
2 g/l	2.5 g/l	333.3333	183.18176	.717	-315.3330	981.9997
	3 g/l	333.3333	183.18176	.717	-315.3330	981.9997
	3.5 g/l	733.3333*	183.18176	.019	84.6670	1381.9997
	4 g/l	766.6667*	183.18176	.013	118.0003	1415.3330
	4.5 g/l	433.3333	183.18176	.393	-215.3330	1081.9997
	0 g/l	-1033.3333*	183.18176	.001	-1681.9997	-384.6670
	0.5 g/l	-1033.3333*	183.18176	.001	-1681.9997	-384.6670
2.5 g/l	1 g/l	-1500.0000*	183.18176	.000	-2148.6663	-851.3337
	1.5 g/l	-866.6667*	183.18176	.004	-1515.3330	-218.0003
	2 g/l	-333.3333	183.18176	.717	-981.9997	315.3330

	3 g/l	.0000	183.18176	1.000	-648.6663	648.6663
	3.5 g/l	400.0000	183.18176	.497	-248.6663	1048.6663
	4 g/l	433.3333	183.18176	.393	-215.3330	1081.9997
	4.5 g/l	100.0000	183.18176	1.000	-548.6663	748.6663
	0 g/l	-1033.3333*	183.18176	.001	-1681.9997	-384.6670
	0.5 g/l	-1033.3333*	183.18176	.001	-1681.9997	-384.6670
	1 g/l	-1500.0000*	183.18176	.000	-2148.6663	-851.3337
	1.5 g/l	-866.6667*	183.18176	.004	-1515.3330	-218.0003
3 g/l	2 g/l	-333.3333	183.18176	.717	-981.9997	315.3330
	2.5 g/l	.0000	183.18176	1.000	-648.6663	648.6663
	3.5 g/l	400.0000	183.18176	.497	-248.6663	1048.6663
	4 g/l	433.3333	183.18176	.393	-215.3330	1081.9997
	4.5 g/l	100.0000	183.18176	1.000	-548.6663	748.6663
	0 g/l	-1433.3333*	183.18176	.000	-2081.9997	-784.6670
	0.5 g/l	-1433.3333*	183.18176	.000	-2081.9997	-784.6670
	1 g/l	-1900.0000*	183.18176	.000	-2548.6663	-1251.3337
	1.5 g/l	-1266.6667*	183.18176	.000	-1915.3330	-618.0003
3.5 g/l	2 g/l	-733.3333*	183.18176	.019	-1381.9997	-84.6670
	2.5 g/l	-400.0000	183.18176	.497	-1048.6663	248.6663
	3 g/l	-400.0000	183.18176	.497	-1048.6663	248.6663
	4 g/l	33.3333	183.18176	1.000	-615.3330	681.9997
	4.5 g/l	-300.0000	183.18176	.815	-948.6663	348.6663
	0 g/l	-1466.6667*	183.18176	.000	-2115.3330	-818.0003
	0.5 g/l	-1466.6667*	183.18176	.000	-2115.3330	-818.0003
	1 g/l	-1933.3333*	183.18176	.000	-2581.9997	-1284.6670
	1.5 g/l	-1300.0000*	183.18176	.000	-1948.6663	-651.3337
4 g/l	2 g/l	-766.6667*	183.18176	.013	-1415.3330	-118.0003
	2.5 g/l	-433.3333	183.18176	.393	-1081.9997	215.3330
	3 g/l	-433.3333	183.18176	.393	-1081.9997	215.3330
	3.5 g/l	-33.3333	183.18176	1.000	-681.9997	615.3330
	4.5 g/l	-333.3333	183.18176	.717	-981.9997	315.3330
	0 g/l	-1133.3333*	183.18176	.000	-1781.9997	-484.6670
	0.5 g/l	-1133.3333*	183.18176	.000	-1781.9997	-484.6670
	1 g/l	-1600.0000*	183.18176	.000	-2248.6663	-951.3337
4.5 g/l	1.5 g/l	-966.6667*	183.18176	.001	-1615.3330	-318.0003
	2 g/l	-433.3333	183.18176	.393	-1081.9997	215.3330
	2.5 g/l	-100.0000	183.18176	1.000	-748.6663	548.6663
	3 g/l	-100.0000	183.18176	1.000	-748.6663	548.6663

300

Biomass		3.5 g/l	300.0000	183.18176	.815	-348.6663	948.6663
		4 g/l	333.3333	183.18176	.717	-315.3330	981.9997
	0 g/l	0.5 g/l	-.3960	.14595	.233	-.9128	.1208
		1 g/l	-.4097	.14595	.199	-.9265	.1071
		1.5 g/l	-.1150	.14595	.998	-.6318	.4018
		2 g/l	.1983	.14595	.926	-.3185	.7151
		2.5 g/l	.2637	.14595	.724	-.2531	.7805
		3 g/l	.2863	.14595	.632	-.2305	.8031
		3.5 g/l	.3440	.14595	.398	-.1728	.8608
		4 g/l	.4183	.14595	.180	-.0985	.9351
		4.5 g/l	.3277	.14595	.461	-.1891	.8445
	0.5 g/l	0 g/l	.3960	.14595	.233	-.1208	.9128
		1 g/l	-.0137	.14595	1.000	-.5305	.5031
		1.5 g/l	.2810	.14595	.654	-.2358	.7978
		2 g/l	.5943*	.14595	.016	.0775	1.1111
		2.5 g/l	.6597*	.14595	.006	.1429	1.1765
		3 g/l	.6823*	.14595	.004	.1655	1.1991
		3.5 g/l	.7400*	.14595	.002	.2232	1.2568
		4 g/l	.8143*	.14595	.001	.2975	1.3311
		4.5 g/l	.7237*	.14595	.002	.2069	1.2405
	1 g/l	0 g/l	.4097	.14595	.199	-.1071	.9265
		0.5 g/l	.0137	.14595	1.000	-.5031	.5305
		1.5 g/l	.2947	.14595	.597	-.2221	.8115
		2 g/l	.6080*	.14595	.013	.0912	1.1248
		2.5 g/l	.6733*	.14595	.005	.1565	1.1901
		3 g/l	.6960*	.14595	.004	.1792	1.2128
		3.5 g/l	.7537*	.14595	.002	.2369	1.2705
		4 g/l	.8280*	.14595	.001	.3112	1.3448
		4.5 g/l	.7373*	.14595	.002	.2205	1.2541
	1.5 g/l	0 g/l	.1150	.14595	.998	-.4018	.6318
		0.5 g/l	-.2810	.14595	.654	-.7978	.2358
		1 g/l	-.2947	.14595	.597	-.8115	.2221
		2 g/l	.3133	.14595	.519	-.2035	.8301
		2.5 g/l	.3787	.14595	.281	-.1381	.8955
		3 g/l	.4013	.14595	.219	-.1155	.9181
		3.5 g/l	.4590	.14595	.109	-.0578	.9758
		4 g/l	.5333*	.14595	.040	.0165	1.0501
		4.5 g/l	.4427	.14595	.134	-.0741	.9595
	2 g/l	0 g/l	-.1983	.14595	.926	-.7151	.3185
		0.5 g/l	-.5943*	.14595	.016	-1.1111	-.0775
		1 g/l	-.6080*	.14595	.013	-1.1248	-.0912
		1.5 g/l	-.3133	.14595	.519	-.8301	.2035
		2.5 g/l	.0653	.14595	1.000	-.4515	.5821

	3 g/l	.0880	.14595	1.000	-.4288	.6048
	3.5 g/l	.1457	.14595	.989	-.3711	.6625
	4 g/l	.2200	.14595	.874	-.2968	.7368
	4.5 g/l	.1293	.14595	.995	-.3875	.6461
	0 g/l	-.2637	.14595	.724	-.7805	.2531
	0.5 g/l	-.6597*	.14595	.006	-1.1765	-.1429
	1 g/l	-.6733*	.14595	.005	-1.1901	-.1565
	1.5 g/l	-.3787	.14595	.281	-.8955	.1381
2.5 g/l	2 g/l	-.0653	.14595	1.000	-.5821	.4515
	3 g/l	.0227	.14595	1.000	-.4941	.5395
	3.5 g/l	.0803	.14595	1.000	-.4365	.5971
	4 g/l	.1547	.14595	.984	-.3621	.6715
	4.5 g/l	.0640	.14595	1.000	-.4528	.5808
	0 g/l	-.2863	.14595	.632	-.8031	.2305
	0.5 g/l	-.6823*	.14595	.004	-1.1991	-.1655
	1 g/l	-.6960*	.14595	.004	-1.2128	-.1792
	1.5 g/l	-.4013	.14595	.219	-.9181	.1155
3 g/l	2 g/l	-.0880	.14595	1.000	-.6048	.4288
	2.5 g/l	-.0227	.14595	1.000	-.5395	.4941
	3.5 g/l	.0577	.14595	1.000	-.4591	.5745
	4 g/l	.1320	.14595	.994	-.3848	.6488
	4.5 g/l	.0413	.14595	1.000	-.4755	.5581
	0 g/l	-.3440	.14595	.398	-.8608	.1728
	0.5 g/l	-.7400*	.14595	.002	-1.2568	-.2232
	1 g/l	-.7537*	.14595	.002	-1.2705	-.2369
	1.5 g/l	-.4590	.14595	.109	-.9758	.0578
3.5 g/l	2 g/l	-.1457	.14595	.989	-.6625	.3711
	2.5 g/l	-.0803	.14595	1.000	-.5971	.4365
	3 g/l	-.0577	.14595	1.000	-.5745	.4591
	4 g/l	.0743	.14595	1.000	-.4425	.5911
	4.5 g/l	-.0163	.14595	1.000	-.5331	.5005
	0 g/l	-.4183	.14595	.180	-.9351	.0985
	0.5 g/l	-.8143*	.14595	.001	-1.3311	-.2975
	1 g/l	-.8280*	.14595	.001	-1.3448	-.3112
	1.5 g/l	-.5333*	.14595	.040	-1.0501	-.0165
4 g/l	2 g/l	-.2200	.14595	.874	-.7368	.2968
	2.5 g/l	-.1547	.14595	.984	-.6715	.3621
	3 g/l	-.1320	.14595	.994	-.6488	.3848
	3.5 g/l	-.0743	.14595	1.000	-.5911	.4425
	4.5 g/l	-.0907	.14595	1.000	-.6075	.4261
	0 g/l	-.3277	.14595	.461	-.8445	.1891
4.5 g/l	0.5 g/l	-.7237*	.14595	.002	-1.2405	-.2069
	1 g/l	-.7373*	.14595	.002	-1.2541	-.2205

	(I)	(J)	Mean Difference	Std. Error	Sig.	Lower Bound	Upper Bound
pH		1.5 g/l	-.4427	.14595	.134	-.9595	.0741
		2 g/l	-.1293	.14595	.995	-.6461	.3875
		2.5 g/l	-.0640	.14595	1.000	-.5808	.4528
		3 g/l	-.0413	.14595	1.000	-.5581	.4755
		3.5 g/l	.0163	.14595	1.000	-.5005	.5331
		4 g/l	.0907	.14595	1.000	-.4261	.6075
		0.5 g/l	.0633	.02028	.113	-.0085	.1351
		1 g/l	.0300	.02028	.885	-.0418	.1018
		1.5 g/l	.0167	.02028	.997	-.0551	.0885
		2 g/l	-.0100	.02028	1.000	-.0818	.0618
	0 g/l	2.5 g/l	-.0667	.02028	.083	-.1385	.0051
		3 g/l	-.0667	.02028	.083	-.1385	.0051
		3.5 g/l	-.0800*	.02028	.022	-.1518	-.0082
		4 g/l	-.1267*	.02028	.000	-.1985	-.0549
		4.5 g/l	-.0933*	.02028	.005	-.1651	-.0215
		0 g/l	-.0633	.02028	.113	-.1351	.0085
		1 g/l	-.0333	.02028	.812	-.1051	.0385
		1.5 g/l	-.0467	.02028	.429	-.1185	.0251
		2 g/l	-.0733*	.02028	.043	-.1451	-.0015
	0.5 g/l	2.5 g/l	-.1300*	.02028	.000	-.2018	-.0582
		3 g/l	-.1300*	.02028	.000	-.2018	-.0582
		3.5 g/l	-.1433*	.02028	.000	-.2151	-.0715
		4 g/l	-.1900*	.02028	.000	-.2618	-.1182
		4.5 g/l	-.1567*	.02028	.000	-.2285	-.0849
		0 g/l	-.0300	.02028	.885	-.1018	.0418
		0.5 g/l	.0333	.02028	.812	-.0385	.1051
		1.5 g/l	-.0133	.02028	1.000	-.0851	.0585
		2 g/l	-.0400	.02028	.625	-.1118	.0318
	1 g/l	2.5 g/l	-.0967*	.02028	.004	-.1685	-.0249
		3 g/l	-.0967*	.02028	.004	-.1685	-.0249
		3.5 g/l	-.1100*	.02028	.001	-.1818	-.0382
		4 g/l	-.1567*	.02028	.000	-.2285	-.0849
		4.5 g/l	-.1233*	.02028	.000	-.1951	-.0515
		0 g/l	-.0167	.02028	.997	-.0885	.0551
		0.5 g/l	.0467	.02028	.429	-.0251	.1185
		1 g/l	.0133	.02028	1.000	-.0585	.0851
		2 g/l	-.0267	.02028	.938	-.0985	.0451
	1.5 g/l	2.5 g/l	-.0833*	.02028	.015	-.1551	-.0115
		3 g/l	-.0833*	.02028	.015	-.1551	-.0115
		3.5 g/l	-.0967*	.02028	.004	-.1685	-.0249
		4 g/l	-.1433*	.02028	.000	-.2151	-.0715
		4.5 g/l	-.1100*	.02028	.001	-.1818	-.0382
	2 g/l	0 g/l	.0100	.02028	1.000	-.0618	.0818

	0.5 g/l	.0733*	.02028	.043	.0015	.1451
	1 g/l	.0400	.02028	.625	-.0318	.1118
	1.5 g/l	.0267	.02028	.938	-.0451	.0985
	2.5 g/l	-.0567	.02028	.203	-.1285	.0151
	3 g/l	-.0567	.02028	.203	-.1285	.0151
	3.5 g/l	-.0700	.02028	.060	-.1418	.0018
	4 g/l	-.1167*	.02028	.000	-.1885	-.0449
	4.5 g/l	-.0833*	.02028	.015	-.1551	-.0115
	0 g/l	.0667	.02028	.083	-.0051	.1385
	0.5 g/l	.1300*	.02028	.000	.0582	.2018
	1 g/l	.0967*	.02028	.004	.0249	.1685
	1.5 g/l	.0833*	.02028	.015	.0115	.1551
2.5 g/l	2 g/l	.0567	.02028	.203	-.0151	.1285
	3 g/l	.0000	.02028	1.000	-.0718	.0718
	3.5 g/l	-.0133	.02028	1.000	-.0851	.0585
	4 g/l	-.0600	.02028	.153	-.1318	.0118
	4.5 g/l	-.0267	.02028	.938	-.0985	.0451
	0 g/l	.0667	.02028	.083	-.0051	.1385
	0.5 g/l	.1300*	.02028	.000	.0582	.2018
	1 g/l	.0967*	.02028	.004	.0249	.1685
	1.5 g/l	.0833*	.02028	.015	.0115	.1551
3 g/l	2 g/l	.0567	.02028	.203	-.0151	.1285
	2.5 g/l	.0000	.02028	1.000	-.0718	.0718
	3.5 g/l	-.0133	.02028	1.000	-.0851	.0585
	4 g/l	-.0600	.02028	.153	-.1318	.0118
	4.5 g/l	-.0267	.02028	.938	-.0985	.0451
	0 g/l	.0800*	.02028	.022	.0082	.1518
	0.5 g/l	.1433*	.02028	.000	.0715	.2151
	1 g/l	.1100*	.02028	.001	.0382	.1818
	1.5 g/l	.0967*	.02028	.004	.0249	.1685
3.5 g/l	2 g/l	.0700	.02028	.060	-.0018	.1418
	2.5 g/l	.0133	.02028	1.000	-.0585	.0851
	3 g/l	.0133	.02028	1.000	-.0585	.0851
	4 g/l	-.0467	.02028	.429	-.1185	.0251
	4.5 g/l	-.0133	.02028	1.000	-.0851	.0585
	0 g/l	.1267*	.02028	.000	.0549	.1985
	0.5 g/l	.1900*	.02028	.000	.1182	.2618
	1 g/l	.1567*	.02028	.000	.0849	.2285
	1.5 g/l	.1433*	.02028	.000	.0715	.2151
4 g/l	2 g/l	.1167*	.02028	.000	.0449	.1885
	2.5 g/l	.0600	.02028	.153	-.0118	.1318
	3 g/l	.0600	.02028	.153	-.0118	.1318
	3.5 g/l	.0467	.02028	.429	-.0251	.1185

	(I)	(J)	Mean Difference (I-J)	Std. Error	Sig.	Lower Bound	Upper Bound
Acetate	4.5 g/l	4.5 g/l	.0333	.02028	.812	-.0385	.1051
		0 g/l	.0933*	.02028	.005	.0215	.1651
		0.5 g/l	.1567*	.02028	.000	.0849	.2285
		1 g/l	.1233*	.02028	.000	.0515	.1951
		1.5 g/l	.1100*	.02028	.001	.0382	.1818
		2 g/l	.0833*	.02028	.015	.0115	.1551
		2.5 g/l	.0267	.02028	.938	-.0451	.0985
		3 g/l	.0267	.02028	.938	-.0451	.0985
		3.5 g/l	.0133	.02028	1.000	-.0585	.0851
		4 g/l	-.0333	.02028	.812	-.1051	.0385
	0 g/l	0.5 g/l	.1042	.08123	.946	-.1834	.3919
		1 g/l	-.0271	.08123	1.000	-.3147	.2606
		1.5 g/l	-.1122	.08123	.919	-.3999	.1754
		2 g/l	.1566	.08123	.653	-.1311	.4442
		2.5 g/l	.4239*	.08123	.001	.1362	.7115
		3 g/l	.6898*	.08123	.000	.4021	.9774
		3.5 g/l	.6366*	.08123	.000	.3490	.9243
		4 g/l	.9670*	.08123	.000	.6794	1.2547
		4.5 g/l	.5720*	.08123	.000	.2844	.8597
	0.5 g/l	0 g/l	-.1042	.08123	.946	-.3919	.1834
		1 g/l	-.1313	.08123	.826	-.4189	.1564
		1.5 g/l	-.2164	.08123	.252	-.5041	.0712
		2 g/l	.0524	.08123	1.000	-.2353	.3400
		2.5 g/l	.3197*	.08123	.022	.0320	.6073
		3 g/l	.5855*	.08123	.000	.2979	.8732
		3.5 g/l	.5324*	.08123	.000	.2448	.8201
		4 g/l	.8628*	.08123	.000	.5751	1.1504
		4.5 g/l	.4678*	.08123	.000	.1801	.7554
	1 g/l	0 g/l	.0271	.08123	1.000	-.2606	.3147
		0.5 g/l	.1313	.08123	.826	-.1564	.4189
		1.5 g/l	-.0852	.08123	.985	-.3728	.2025
		2 g/l	.1837	.08123	.452	-.1040	.4713
		2.5 g/l	.4510*	.08123	.001	.1633	.7386
		3 g/l	.7168*	.08123	.000	.4292	1.0045
		3.5 g/l	.6637*	.08123	.000	.3761	.9514
		4 g/l	.9941*	.08123	.000	.7064	1.2817
		4.5 g/l	.5991*	.08123	.000	.3114	.8867
	1.5 g/l	0 g/l	.1122	.08123	.919	-.1754	.3999
		0.5 g/l	.2164	.08123	.252	-.0712	.5041
		1 g/l	.0852	.08123	.985	-.2025	.3728
		2 g/l	.2688	.08123	.079	-.0188	.5565
		2.5 g/l	.5361*	.08123	.000	.2485	.8238
		3 g/l	.8020*	.08123	.000	.5143	1.0896

		3.5 g/l	.7489*	.08123	.000	.4612	1.0365
		4 g/l	1.0792*	.08123	.000	.7916	1.3669
		4.5 g/l	.6842*	.08123	.000	.3966	.9719
		0 g/l	-.1566	.08123	.653	-.4442	.1311
		0.5 g/l	-.0524	.08123	1.000	-.3400	.2353
		1 g/l	-.1837	.08123	.452	-.4713	.1040
		1.5 g/l	-.2688	.08123	.079	-.5565	.0188
	2 g/l	2.5 g/l	.2673	.08123	.082	-.0204	.5549
		3 g/l	.5332*	.08123	.000	.2455	.8208
		3.5 g/l	.4801*	.08123	.000	.1924	.7677
		4 g/l	.8104*	.08123	.000	.5228	1.0981
		4.5 g/l	.4154*	.08123	.002	.1278	.7031
		0 g/l	-.4239*	.08123	.001	-.7115	-.1362
		0.5 g/l	-.3197*	.08123	.022	-.6073	-.0320
		1 g/l	-.4510*	.08123	.001	-.7386	-.1633
		1.5 g/l	-.5361*	.08123	.000	-.8238	-.2485
	2.5 g/l	2 g/l	-.2673	.08123	.082	-.5549	.0204
		3 g/l	.2659	.08123	.085	-.0218	.5535
		3.5 g/l	.2128	.08123	.271	-.0749	.5004
		4 g/l	.5431*	.08123	.000	.2555	.8308
		4.5 g/l	.1481	.08123	.715	-.1395	.4358
		0 g/l	-.6898*	.08123	.000	-.9774	-.4021
		0.5 g/l	-.5855*	.08123	.000	-.8732	-.2979
		1 g/l	-.7168*	.08123	.000	-1.0045	-.4292
		1.5 g/l	-.8020*	.08123	.000	-1.0896	-.5143
	3 g/l	2 g/l	-.5332*	.08123	.000	-.8208	-.2455
		2.5 g/l	-.2659	.08123	.085	-.5535	.0218
		3.5 g/l	-.0531	.08123	1.000	-.3408	.2345
		4 g/l	.2773	.08123	.065	-.0104	.5649
		4.5 g/l	-.1177	.08123	.896	-.4054	.1699
		0 g/l	-.6366*	.08123	.000	-.9243	-.3490
		0.5 g/l	-.5324*	.08123	.000	-.8201	-.2448
		1 g/l	-.6637*	.08123	.000	-.9514	-.3761
		1.5 g/l	-.7489*	.08123	.000	-1.0365	-.4612
	3.5 g/l	2 g/l	-.4801*	.08123	.000	-.7677	-.1924
		2.5 g/l	-.2128	.08123	.271	-.5004	.0749
		3 g/l	.0531	.08123	1.000	-.2345	.3408
		4 g/l	.3304*	.08123	.017	.0427	.6180
		4.5 g/l	-.0646	.08123	.998	-.3523	.2230
		0 g/l	-.9670*	.08123	.000	-1.2547	-.6794
	4 g/l	0.5 g/l	-.8628*	.08123	.000	-1.1504	-.5751
		1 g/l	-.9941*	.08123	.000	-1.2817	-.7064
		1.5 g/l	-1.0792*	.08123	.000	-1.3669	-.7916

Butyrate							
		2 g/l	-.8104*	.08123	.000	-1.0981	-.5228
		2.5 g/l	-.5431*	.08123	.000	-.8308	-.2555
		3 g/l	-.2773	.08123	.065	-.5649	.0104
		3.5 g/l	-.3304*	.08123	.017	-.6180	-.0427
		4.5 g/l	-.3950*	.08123	.003	-.6826	-.1073
		0 g/l	-.5720*	.08123	.000	-.8597	-.2844
		0.5 g/l	-.4678*	.08123	.000	-.7554	-.1801
		1 g/l	-.5991*	.08123	.000	-.8867	-.3114
		1.5 g/l	-.6842*	.08123	.000	-.9719	-.3966
	4.5 g/l	2 g/l	-.4154*	.08123	.002	-.7031	-.1278
		2.5 g/l	-.1481	.08123	.715	-.4358	.1395
		3 g/l	.1177	.08123	.896	-.1699	.4054
		3.5 g/l	.0646	.08123	.998	-.2230	.3523
		4 g/l	.3950*	.08123	.003	.1073	.6826
		0.5 g/l	-.2249	.07362	.128	-.4857	.0358
		1 g/l	-.3611*	.07362	.003	-.6218	-.1004
		1.5 g/l	-.3635*	.07362	.003	-.6242	-.1028
		2 g/l	-.6029*	.07362	.000	-.8636	-.3422
	0 g/l	2.5 g/l	-.8614*	.07362	.000	-1.1221	-.6007
		3 g/l	-1.0478*	.07362	.000	-1.3085	-.7871
		3.5 g/l	-1.0745*	.07362	.000	-1.3352	-.8138
		4 g/l	-.5506*	.07362	.000	-.8113	-.2899
		4.5 g/l	-.8457*	.07362	.000	-1.1064	-.5850
		0 g/l	.2249	.07362	.128	-.0358	.4857
		1 g/l	-.1362	.07362	.699	-.3969	.1245
		1.5 g/l	-.1386	.07362	.680	-.3993	.1221
		2 g/l	-.3779*	.07362	.002	-.6386	-.1172
	0.5 g/l	2.5 g/l	-.6365*	.07362	.000	-.8972	-.3758
		3 g/l	-.8228*	.07362	.000	-1.0835	-.5621
		3.5 g/l	-.8496*	.07362	.000	-1.1103	-.5889
		4 g/l	-.3257*	.07362	.008	-.5864	-.0650
		4.5 g/l	-.6207*	.07362	.000	-.8814	-.3600
		0 g/l	.3611*	.07362	.003	.1004	.6218
		0.5 g/l	.1362	.07362	.699	-.1245	.3969
		1.5 g/l	-.0024	.07362	1.000	-.2631	.2583
		2 g/l	-.2418	.07362	.083	-.5025	.0190
	1 g/l	2.5 g/l	-.5003*	.07362	.000	-.7610	-.2396
		3 g/l	-.6866*	.07362	.000	-.9473	-.4259
		3.5 g/l	-.7134*	.07362	.000	-.9741	-.4527
		4 g/l	-.1895	.07362	.291	-.4502	.0712

307

		4.5 g/l	-.4845*	.07362	.000	-.7452	-.2238
		0 g/l	.3635*	.07362	.003	.1028	.6242
		0.5 g/l	.1386	.07362	.680	-.1221	.3993
		1 g/l	.0024	.07362	1.000	-.2583	.2631
		2 g/l	-.2394	.07362	.089	-.5001	.0213
1.5 g/l		2.5 g/l	-.4979*	.07362	.000	-.7586	-.2372
		3 g/l	-.6842*	.07362	.000	-.9449	-.4235
		3.5 g/l	-.7110*	.07362	.000	-.9717	-.4503
		4 g/l	-.1871	.07362	.305	-.4478	.0736
		4.5 g/l	-.4821*	.07362	.000	-.7429	-.2214
		0 g/l	.6029*	.07362	.000	.3422	.8636
		0.5 g/l	.3779*	.07362	.002	.1172	.6386
		1 g/l	.2418	.07362	.083	-.0190	.5025
		1.5 g/l	.2394	.07362	.089	-.0213	.5001
2 g/l		2.5 g/l	-.2585	.07362	.053	-.5192	.0022
		3 g/l	-.4449*	.07362	.000	-.7056	-.1842
		3.5 g/l	-.4716*	.07362	.000	-.7324	-.2109
		4 g/l	.0523	.07362	.999	-.2084	.3130
		4.5 g/l	-.2428	.07362	.081	-.5035	.0179
		0 g/l	.8614*	.07362	.000	.6007	1.1221
		0.5 g/l	.6365*	.07362	.000	.3758	.8972
		1 g/l	.5003*	.07362	.000	.2396	.7610
		1.5 g/l	.4979*	.07362	.000	.2372	.7586
2.5 g/l		2 g/l	.2585	.07362	.053	-.0022	.5192
		3 g/l	-.1863	.07362	.310	-.4470	.0744
		3.5 g/l	-.2131	.07362	.171	-.4738	.0476
		4 g/l	.3108*	.07362	.012	.0501	.5715
		4.5 g/l	.0157	.07362	1.000	-.2450	.2765
		0 g/l	1.0478*	.07362	.000	.7871	1.3085
		0.5 g/l	.8228*	.07362	.000	.5621	1.0835
		1 g/l	.6866*	.07362	.000	.4259	.9473
		1.5 g/l	.6842*	.07362	.000	.4235	.9449
3 g/l		2 g/l	.4449*	.07362	.000	.1842	.7056
		2.5 g/l	.1863	.07362	.310	-.0744	.4470
		3.5 g/l	-.0268	.07362	1.000	-.2875	.2339
		4 g/l	.4971*	.07362	.000	.2364	.7579
		4.5 g/l	.2021	.07362	.221	-.0586	.4628
3.5 g/l		0 g/l	1.0745*	.07362	.000	.8138	1.3352

	0.5 g/l	.8496*	.07362	.000	.5889	1.1103
	1 g/l	.7134*	.07362	.000	.4527	.9741
	1.5 g/l	.7110*	.07362	.000	.4503	.9717
	2 g/l	.4716*	.07362	.000	.2109	.7324
	2.5 g/l	.2131	.07362	.171	-.0476	.4738
	3 g/l	.0268	.07362	1.000	-.2339	.2875
	4 g/l	.5239*	.07362	.000	.2632	.7846
	4.5 g/l	.2289	.07362	.116	-.0318	.4896
	0 g/l	.5506*	.07362	.000	.2899	.8113
	0.5 g/l	.3257*	.07362	.008	.0650	.5864
	1 g/l	.1895	.07362	.291	-.0712	.4502
	1.5 g/l	.1871	.07362	.305	-.0736	.4478
4 g/l	2 g/l	-.0523	.07362	.999	-.3130	.2084
	2.5 g/l	-.3108*	.07362	.012	-.5715	-.0501
	3 g/l	-.4971*	.07362	.000	-.7579	-.2364
	3.5 g/l	-.5239*	.07362	.000	-.7846	-.2632
	4.5 g/l	-.2951*	.07362	.019	-.5558	-.0344
	0 g/l	.8457*	.07362	.000	.5850	1.1064
	0.5 g/l	.6207*	.07362	.000	.3600	.8814
	1 g/l	.4845*	.07362	.000	.2238	.7452
	1.5 g/l	.4821*	.07362	.000	.2214	.7429
4.5 g/l	2 g/l	.2428	.07362	.081	-.0179	.5035
	2.5 g/l	-.0157	.07362	1.000	-.2765	.2450
	3 g/l	-.2021	.07362	.221	-.4628	.0586
	3.5 g/l	-.2289	.07362	.116	-.4896	.0318
	4 g/l	.2951*	.07362	.019	.0344	.5558

Based on observed means.

The error term is Mean Square(Error) = .008.

*. The mean difference is significant at the .05 level.

Table 4.6 Statistical analysis between independent variable L-cysteine.HCl.H₂O and dependent variables mole of acetate and butyrate produced per N-acetylglucosamine utilized.

Source	Dependent Variable	Sig.
Corrected Model	Acetate mol per mol GlcNAc	.000[a]
	Butyrate mol per mol GlcNAc	.000[b]
Intercept	Acetate mol per mol GlcNAc	.000
	Butyrate mol per mol GlcNAc	.000
Cysteine	Acetate mol per mol GlcNAc	.000
	Butyrate mol per mol GlcNAc	.000
Error	Acetate mol per mol GlcNAc	
	Butyrate mol per mol GlcNAc	
Total	Acetate mol per mol GlcNAc	
	Butyrate mol per mol GlcNAc	
Corrected Total	Acetate mol per mol GlcNAc	
	Butyrate mol per mol GlcNAc	

a. R Squared = .913 (Adjusted R Squared = .874)

b. R Squared = .967 (Adjusted R Squared = .952)

Table 4.7 Post Hoc, Tukey HSD multiple comparison statistical analysis between independent variable L-cysteine.HCl.H$_2$O and dependent variables mole of acetate and butyrate produced per N-acetylglucosamine utilized.

Dependent Variable	(I) Cysteine	(J) Cysteine	Mean Difference (I-J)	Std. Error	Sig.	95% Confidence Interval	
						Lower Bound	Upper Bound
Acetate mol per mol GlcNAc	0 g/l	0.5	.0128	.02114	1.000	-.0621	.0876
		1 g/l	-.0347	.02114	.814	-.1095	.0402
		1.5 g/l	-.0495	.02114	.406	-.1244	.0254
		2 g/l	.0385	.02114	.716	-.0364	.1133
		2.5 g/l	.0971*	.02114	.005	.0222	.1720
		3 g/l	.1364*	.02114	.000	.0615	.2112
		3.5 g/l	.1442*	.02114	.000	.0693	.2190
		4 g/l	.1231*	.02114	.000	.0483	.1980
		4.5 g/l	.0975*	.02114	.005	.0226	.1724
	0.5	0 g/l	-.0128	.02114	1.000	-.0876	.0621
		1 g/l	-.0474	.02114	.461	-.1223	.0274
		1.5 g/l	-.0623	.02114	.156	-.1372	.0126
		2 g/l	.0257	.02114	.961	-.0492	.1006
		2.5 g/l	.0843*	.02114	.020	.0095	.1592
		3 g/l	.1236*	.02114	.000	.0487	.1985

310

		3.5 g/l	.1314[*]	.02114	.000	.0565	.2062
		4 g/l	.1104[*]	.02114	.001	.0355	.1852
		4.5 g/l	.0847[*]	.02114	.019	.0098	.1596
		0 g/l	.0347	.02114	.814	-.0402	.1095
		0.5	.0474	.02114	.461	-.0274	.1223
		1.5 g/l	-.0149	.02114	.999	-.0897	.0600
		2 g/l	.0731	.02114	.059	-.0017	.1480
1 g/l		2.5 g/l	.1318[*]	.02114	.000	.0569	.2066
		3 g/l	.1710[*]	.02114	.000	.0962	.2459
		3.5 g/l	.1788[*]	.02114	.000	.1040	.2537
		4 g/l	.1578[*]	.02114	.000	.0829	.2327
		4.5 g/l	.1322[*]	.02114	.000	.0573	.2070
		0 g/l	.0495	.02114	.406	-.0254	.1244
		0.5	.0623	.02114	.156	-.0126	.1372
		1 g/l	.0149	.02114	.999	-.0600	.0897
		2 g/l	.0880[*]	.02114	.014	.0131	.1629
1.5 g/l		2.5 g/l	.1466[*]	.02114	.000	.0718	.2215
		3 g/l	.1859[*]	.02114	.000	.1110	.2608
		3.5 g/l	.1937[*]	.02114	.000	.1188	.2685
		4 g/l	.1727[*]	.02114	.000	.0978	.2475
		4.5 g/l	.1470[*]	.02114	.000	.0721	.2219
		0 g/l	-.0385	.02114	.716	-.1133	.0364
		0.5	-.0257	.02114	.961	-.1006	.0492
		1 g/l	-.0731	.02114	.059	-.1480	.0017
		1.5 g/l	-.0880[*]	.02114	.014	-.1629	-.0131
2 g/l		2.5 g/l	.0586	.02114	.211	-.0162	.1335
		3 g/l	.0979[*]	.02114	.005	.0230	.1728
		3.5 g/l	.1057[*]	.02114	.002	.0308	.1805
		4 g/l	.0847[*]	.02114	.019	.0098	.1595
		4.5 g/l	.0590	.02114	.205	-.0159	.1339
		0 g/l	-.0971[*]	.02114	.005	-.1720	-.0222
		0.5	-.0843[*]	.02114	.020	-.1592	-.0095
		1 g/l	-.1318[*]	.02114	.000	-.2066	-.0569
		1.5 g/l	-.1466[*]	.02114	.000	-.2215	-.0718
2.5 g/l		2 g/l	-.0586	.02114	.211	-.1335	.0162
		3 g/l	.0393	.02114	.695	-.0356	.1141
		3.5 g/l	.0471	.02114	.472	-.0278	.1219
		4 g/l	.0260	.02114	.958	-.0488	.1009

		4.5 g/l	.0004	.02114	1.000	-.0745	.0752
		0 g/l	-.1364*	.02114	.000	-.2112	-.0615
		0.5	-.1236*	.02114	.000	-.1985	-.0487
		1 g/l	-.1710*	.02114	.000	-.2459	-.0962
		1.5 g/l	-.1859*	.02114	.000	-.2608	-.1110
	3 g/l	2 g/l	-.0979*	.02114	.005	-.1728	-.0230
		2.5 g/l	-.0393	.02114	.695	-.1141	.0356
		3.5 g/l	.0078	.02114	1.000	-.0671	.0827
		4 g/l	-.0132	.02114	1.000	-.0881	.0616
		4.5 g/l	-.0389	.02114	.705	-.1137	.0360
		0 g/l	-.1442*	.02114	.000	-.2190	-.0693
		0.5	-.1314*	.02114	.000	-.2062	-.0565
		1 g/l	-.1788*	.02114	.000	-.2537	-.1040
		1.5 g/l	-.1937*	.02114	.000	-.2685	-.1188
	3.5 g/l	2 g/l	-.1057*	.02114	.002	-.1805	-.0308
		2.5 g/l	-.0471	.02114	.472	-.1219	.0278
		3 g/l	-.0078	.02114	1.000	-.0827	.0671
		4 g/l	-.0210	.02114	.989	-.0959	.0538
		4.5 g/l	-.0467	.02114	.483	-.1215	.0282
		0 g/l	-.1231*	.02114	.000	-.1980	-.0483
		0.5	-.1104*	.02114	.001	-.1852	-.0355
		1 g/l	-.1578*	.02114	.000	-.2327	-.0829
		1.5 g/l	-.1727*	.02114	.000	-.2475	-.0978
	4 g/l	2 g/l	-.0847*	.02114	.019	-.1595	-.0098
		2.5 g/l	-.0260	.02114	.958	-.1009	.0488
		3 g/l	.0132	.02114	1.000	-.0616	.0881
		3.5 g/l	.0210	.02114	.989	-.0538	.0959
		4.5 g/l	-.0257	.02114	.961	-.1005	.0492
		0 g/l	-.0975*	.02114	.005	-.1724	-.0226
		0.5	-.0847*	.02114	.019	-.1596	-.0098
		1 g/l	-.1322*	.02114	.000	-.2070	-.0573
		1.5 g/l	-.1470*	.02114	.000	-.2219	-.0721
	4.5 g/l	2 g/l	-.0590	.02114	.205	-.1339	.0159
		2.5 g/l	-.0004	.02114	1.000	-.0752	.0745
		3 g/l	.0389	.02114	.705	-.0360	.1137
		3.5 g/l	.0467	.02114	.483	-.0282	.1215
		4 g/l	.0257	.02114	.961	-.0492	.1005
Butyrate mol per	0 g/l	0.5	-.0466*	.01283	.041	-.0920	-.0012

312

mol GlcNAc							
		1 g/l	-.0794*	.01283	.000	-.1248	-.0340
		1.5 g/l	-.0753*	.01283	.000	-.1208	-.0299
		2 g/l	-.1020*	.01283	.000	-.1474	-.0566
		2.5 g/l	-.1521*	.01283	.000	-.1975	-.1067
		3 g/l	-.2089*	.01283	.000	-.2543	-.1635
		3.5 g/l	-.1940*	.01283	.000	-.2395	-.1486
		4 g/l	-.1976*	.01283	.000	-.2431	-.1522
		4.5 g/l	-.1808*	.01283	.000	-.2262	-.1353
		0 g/l	.0466*	.01283	.041	.0012	.0920
		1 g/l	-.0328	.01283	.299	-.0782	.0126
		1.5 g/l	-.0287	.01283	.464	-.0741	.0167
		2 g/l	-.0554*	.01283	.010	-.1008	-.0099
	0.5	2.5 g/l	-.1055*	.01283	.000	-.1509	-.0600
		3 g/l	-.1623*	.01283	.000	-.2077	-.1169
		3.5 g/l	-.1474*	.01283	.000	-.1929	-.1020
		4 g/l	-.1510*	.01283	.000	-.1964	-.1056
		4.5 g/l	-.1342*	.01283	.000	-.1796	-.0887
		0 g/l	.0794*	.01283	.000	.0340	.1248
		0.5	.0328	.01283	.299	-.0126	.0782
		1.5 g/l	.0041	.01283	1.000	-.0414	.0495
		2 g/l	-.0226	.01283	.750	-.0680	.0228
	1 g/l	2.5 g/l	-.0727*	.01283	.001	-.1181	-.0273
		3 g/l	-.1295*	.01283	.000	-.1749	-.0841
		3.5 g/l	-.1147*	.01283	.000	-.1601	-.0692
		4 g/l	-.1182*	.01283	.000	-.1637	-.0728
		4.5 g/l	-.1014*	.01283	.000	-.1468	-.0559
		0 g/l	.0753*	.01283	.000	.0299	.1208
		0.5	.0287	.01283	.464	-.0167	.0741
		1 g/l	-.0041	.01283	1.000	-.0495	.0414
		2 g/l	-.0266	.01283	.561	-.0721	.0188
	1.5 g/l	2.5 g/l	-.0767*	.01283	.000	-.1222	-.0313
		3 g/l	-.1336*	.01283	.000	-.1790	-.0881
		3.5 g/l	-.1187*	.01283	.000	-.1641	-.0733
		4 g/l	-.1223*	.01283	.000	-.1677	-.0769
		4.5 g/l	-.1054*	.01283	.000	-.1509	-.0600
		0 g/l	.1020*	.01283	.000	.0566	.1474
	2 g/l	0.5	.0554*	.01283	.010	.0099	.1008
		1 g/l	.0226	.01283	.750	-.0228	.0680

	1.5 g/l	.0266	.01283	.561	-.0188	.0721
	2.5 g/l	-.0501*	.01283	.023	-.0955	-.0047
	3 g/l	-.1069*	.01283	.000	-.1524	-.0615
	3.5 g/l	-.0921*	.01283	.000	-.1375	-.0466
	4 g/l	-.0957*	.01283	.000	-.1411	-.0502
	4.5 g/l	-.0788*	.01283	.000	-.1242	-.0334
	0 g/l	.1521*	.01283	.000	.1067	.1975
	0.5	.1055*	.01283	.000	.0600	.1509
	1 g/l	.0727*	.01283	.001	.0273	.1181
	1.5 g/l	.0767*	.01283	.000	.0313	.1222
2.5 g/l	2 g/l	.0501*	.01283	.023	.0047	.0955
	3 g/l	-.0568*	.01283	.008	-.1023	-.0114
	3.5 g/l	-.0420	.01283	.085	-.0874	.0035
	4 g/l	-.0456*	.01283	.049	-.0910	-.0001
	4.5 g/l	-.0287	.01283	.466	-.0741	.0167
	0 g/l	.2089*	.01283	.000	.1635	.2543
	0.5	.1623*	.01283	.000	.1169	.2077
	1 g/l	.1295*	.01283	.000	.0841	.1749
	1.5 g/l	.1336*	.01283	.000	.0881	.1790
3 g/l	2 g/l	.1069*	.01283	.000	.0615	.1524
	2.5 g/l	.0568*	.01283	.008	.0114	.1023
	3.5 g/l	.0149	.01283	.971	-.0306	.0603
	4 g/l	.0113	.01283	.996	-.0342	.0567
	4.5 g/l	.0281	.01283	.491	-.0173	.0736
	0 g/l	.1940*	.01283	.000	.1486	.2395
	0.5	.1474*	.01283	.000	.1020	.1929
	1 g/l	.1147*	.01283	.000	.0692	.1601
	1.5 g/l	.1187*	.01283	.000	.0733	.1641
3.5 g/l	2 g/l	.0921*	.01283	.000	.0466	.1375
	2.5 g/l	.0420	.01283	.085	-.0035	.0874
	3 g/l	-.0149	.01283	.971	-.0603	.0306
	4 g/l	-.0036	.01283	1.000	-.0490	.0418
	4.5 g/l	.0133	.01283	.986	-.0321	.0587
	0 g/l	.1976*	.01283	.000	.1522	.2431
	0.5	.1510*	.01283	.000	.1056	.1964
4 g/l	1 g/l	.1182*	.01283	.000	.0728	.1637
	1.5 g/l	.1223*	.01283	.000	.0769	.1677
	2 g/l	.0957*	.01283	.000	.0502	.1411

314

	2.5 g/l	.0456*	.01283	.049	.0001	.0910
	3 g/l	-.0113	.01283	.996	-.0567	.0342
	3.5 g/l	.0036	.01283	1.000	-.0418	.0490
	4.5 g/l	.0169	.01283	.938	-.0286	.0623
	0 g/l	.1808*	.01283	.000	.1353	.2262
	0.5	.1342*	.01283	.000	.0887	.1796
	1 g/l	.1014*	.01283	.000	.0559	.1468
	1.5 g/l	.1054*	.01283	.000	.0600	.1509
4.5 g/l	2 g/l	.0788*	.01283	.000	.0334	.1242
	2.5 g/l	.0287	.01283	.466	-.0167	.0741
	3 g/l	-.0281	.01283	.491	-.0736	.0173
	3.5 g/l	-.0133	.01283	.986	-.0587	.0321
	4 g/l	-.0169	.01283	.938	-.0623	.0286

Based on observed means.

The error term is Mean Square(Error) = .000.

*. The mean difference is significant at the .05 level.

Table 4.8 Correlations between final culture pH and final acetate concentration.

		Culture final pH	Acetate
Culture final pH	Pearson Correlation	1	-.873**
	Sig. (2-tailed)		.000
	N	30	30
Acetate	Pearson Correlation	-.873**	1
	Sig. (2-tailed)	.000	
	N	30	30

**. Correlation is significant at the 0.01 level (2-tailed).

315

Table 4.9 Correlations between final culture pH and final butyrate concentration.

		Culture final pH	Butyrate
Culture final pH	Pearson Correlation	1	.623**
	Sig. (2-tailed)		.000
	N	30	30
Butyrate	Pearson Correlation	.623**	1
	Sig. (2-tailed)	.000	
	N	30	30

**. Correlation is significant at the 0.01 level (2-tailed).

Table 4.10 Correlations between Initial L-cysteine.HCl.H$_2$O concentrations and final acetate concentration.

		L-cysteine.HCl.H$_2$O	Acetate
L-cysteine. HCl.H$_2$O	Pearson Correlation	1	-.838**
	Sig. (2-tailed)		.000
	N	30	30
Acetate	Pearson Correlation	-.838**	1
	Sig. (2-tailed)	.000	
	N	30	30

**. Correlation is significant at the 0.01 level (2-tailed).

Table 4.11 Correlations between Initial L-cysteine.HCl.H$_2$O concentrations and final butyrate concentration.

		L-cysteine.HCl.H$_2$O	Butyrate
L-cysteine. HCl.H$_2$O	Pearson Correlation	1	.785**
	Sig. (2-tailed)		.000
	N	30	30
Butyrate	Pearson Correlation	.785**	1
	Sig. (2-tailed)	.000	
	N	30	30

**. Correlation is significant at the 0.01 level (2-tailed).

Table 4.12 Correlations between total biogas volume and final acetate concentration.

		Biogas	Acetate
Biogas	Pearson Correlation	1	.830**
	Sig. (2-tailed)		.000
	N	30	30
Acetate	Pearson Correlation	.830**	1
	Sig. (2-tailed)	.000	
	N	30	30

**. Correlation is significant at the 0.01 level (2-tailed).

Table 4.13 Correlations between Total biogas volume and final butyrate concentration.

		Biogas	Butyrate
Biogas	Pearson Correlation	1	-.730**
	Sig. (2-tailed)		.000
	N	30	30
Butyrate	Pearson Correlation	-.730**	1
	Sig. (2-tailed)	.000	
	N	30	30

**. Correlation is significant at the 0.01 level (2-tailed).

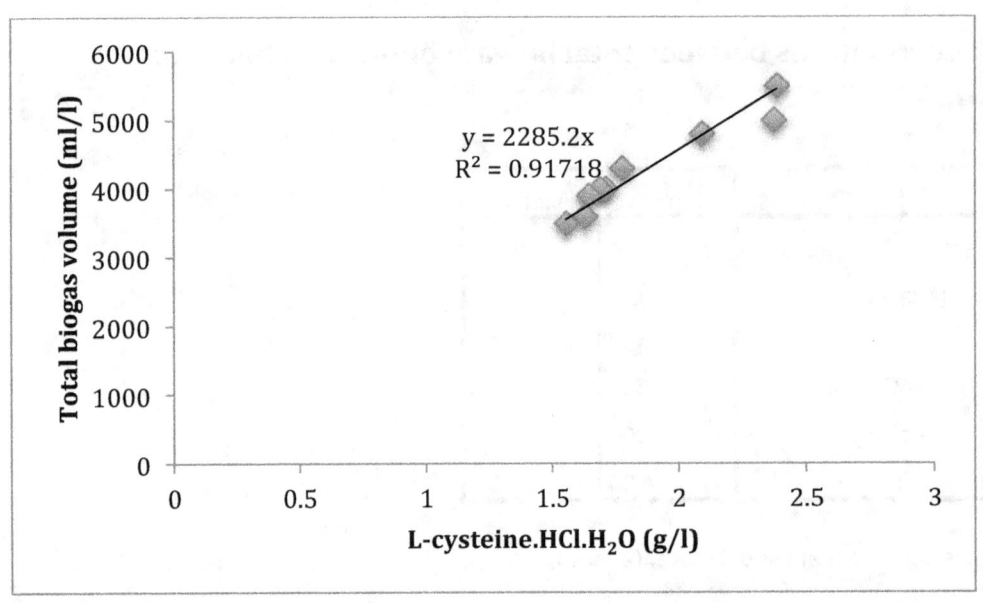

Figure 4.14 Total biogas volume (ml/l) and L-cysteine.HCl.H₂O concentration (g/l) correlation curve.

Appendix 4.2 Theoretical calculation of Hydrogen gas.

Theoretical hydrogen gas is calculated by the total moles of hydrogen gas produced per mole of N-acetylglucosamine utilized.

N-acetylglucosamine = $C_8H_{15}NO_6$ = 15 hydrogen atoms = GlcNAc

Acetete = $C_2H_3O_2^-$ = 3 hydrogen atoms

Butyrate = $C_4H_7O_2^-$ = 7 hydrogen atoms

Hydrogen atomic number = 1= momoatomic

Hydrogen gas = Diatomic = H_2

Theoretical H₂ =

(15-((mole of acetate/mol GlcNAc × 3) + (mole of butyrate/mol GlcNAc × 7)))÷2

Appendix 5

Table 5.1 Total volume of biogas in triplicate for *C. beijerinckii* cultures with N-acetylglucosamine as the main carbon source using the Method of Steepest Ascent design.

Trial Number	Biogas Volume (ml) a	b	c	Biogas Volume Average (ml)	S.D	S.E
1	45	41	38	41.33	3.51	2.03
2	53	60	59	57.33	3.79	2.19
3	56	35	53	48.00	11.36	6.56
4	32	33	32	32.33	0.58	0.33
5	8	10	7	8.33	1.53	0.88

ml = millilitres.
S.D = Standard deviation.
S.E = Standard error.

Table 5.2 Total volume of biogas in triplicate for *C. paraputrificum* cultures with N-acetylglucosamine as the main carbon source using the Method of Steepest Ascent design.

Trial Number	Biogas Volume (ml) a	b	c	Biogas Volume Average (ml)	S.D	S.E
1	21	25	25	23.67	2.31	1.33
2	31	32	36	33.00	2.65	1.53
3	45	41	47	44.33	3.06	1.76
4	54	52	60	55.33	4.16	2.40
5	52	54	54	53.33	1.15	0.67
6	27	28	30	28.33	1.53	0.88

ml = millilitres.
S.D = Standard deviation.
S.E = Standard error.

Table 5.3 Biomass concentrations in triplicate for *C. beijerinckii* cultures with N-acetylglucosamine as the main carbon source using the Method of Steepest Ascent design.

Trial Number	Final O.D			O.D Average	S.D	S.E
	a	b	c			
1	1.976	2.068	2.052	2.032	0.05	0.03
2	3.488	3.444	3.516	3.483	0.04	0.02
3	4.264	3.668	4.264	4.065	0.34	0.20
4	2.812	2.860	2.820	2.831	0.03	0.01
5	1.440	1.492	1.840	1.591	1.64	0.95

O.D = Culture optical density in absorbance unit.
S.D = Standard deviation.
S.E = Standard error.

Table 5.4 Biomass concentrations in triplicate for *C. paraputrificum* cultures with N-acetylglucosamine as the main carbon source using the Method of Steepest Ascent design.

Trial Number	Final O.D			O.D Average	S.D	S.E
	a	b	c			
1	0.714	0.752	0.752	0.739	0.02	0.01
2	1.316	1.424	1.280	1.340	0.07	0.04
3	1.616	1.728	1.730	1.691	0.07	0.04
4	1.770	1.779	1.839	1.796	0.04	0.02
5	2.856	2.334	2.466	2.552	0.27	0.16
6	1.108	0.998	1.119	1.075	0.07	0.04

O.D = Culture optical density in absorbance unit.
S.D = Standard deviation.
S.E = Standard error.

Table 5.5 Culture final pH in triplicate for *C. beijerinckii* cultures with N-acetylglucosamine as the main carbon source using the Method of Steepest Ascent design.

Culture Trials	Final pH			Final pH Average	S.D	S.E
	a	b	c			
1	5.75	5.75	5.74	5.75	0.01	0.00
2	5.48	5.43	5.43	5.45	0.03	0.02
3	5.22	5.23	5.21	5.22	0.01	0.01
4	5.06	5.05	5.02	5.04	0.02	0.01
5	5.13	5.10	5.09	5.11	0.02	0.01

S.D = Standard deviation.
S.E = Standard error.

Table 5.6 Culture final pH in triplicate for *C. paraputrificum* cultures with N-acetylglucosamine as the main carbon source using the Method of Steepest Ascent design.

Culture	Final pH			Final pH		
Trials	a	b	c	Average	S.D	S.E
1	6.05	6.04	6.04	6.04	0.01	0.00
2	5.85	5.81	5.79	5.82	0.03	0.02
3	5.50	5.49	5.49	5.49	0.01	0.00
4	5.33	5.31	5.34	5.33	0.02	0.01
5	5.24	5.26	5.23	5.24	0.02	0.01
6	5.28	5.19	5.20	5.22	0.05	0.03

S.D = Standard deviation.
S.E = Standard error.

Table 5.7 Analysis of variance showing biogas increase in all experimental trials (*C. beijerinckii*).

	Sum of Squares	df	Mean Square	F	Sig.
Between Groups	41870666.667	4	10467666.667	33.056	.000
Within Groups	3166666.667	10	316666.667		
Total	45037333.333	14			

Table 5.8 Analysis of variance showing biogas increase in all experimental trials (*C. paraputrificum*).

	Sum of Squares	df	Mean Square	F	Sig.
Between Groups	26486666.667	5	5297333.333	74.494	.000
Within Groups	853333.333	12	71111.111		
Total	27340000.000	17			

Table 5.9 Post Hoc, Tukey HSD multiple comparison statistical analysis between independent variables in each medium trials and dependent variable total biogas volume (*C. beijerinckii*)

T	(J) Trials	Mean Difference (I-J)	Std. Error	Sig.	95% Confidence Interval	
					Lower Bound	Upper Bound
1	2.00	-1600.0000*	459.46829	.037	-3112.1479	-87.8521
	3.00	-666.66667	459.46829	.612	-2178.8146	845.4812
	4.00	900.00000	459.46829	.349	-612.1479	2412.1479
	5.00	3300.00000*	459.46829	.000	1787.8521	4812.1479
2	1.00	1600.00000*	459.46829	.037	87.8521	3112.1479
	3.00	933.33333	459.46829	.319	-578.8146	2445.4812
	4.00	2500.00000*	459.46829	.002	987.8521	4012.1479
	5.00	4900.00000*	459.46829	.000	3387.8521	6412.1479
3	1.00	666.66667	459.46829	.612	-845.4812	2178.8146
	2.00	-933.33333	459.46829	.319	-2445.4812	578.8146
	4.00	1566.66667*	459.46829	.042	54.5188	3078.8146
	5.00	3966.66667*	459.46829	.000	2454.5188	5478.8146
4	1.00	-900.00000	459.46829	.349	-2412.1479	612.1479
	2.00	-2500.00000*	459.46829	.002	-4012.1479	-987.8521
	3.00	-1566.66667*	459.46829	.042	-3078.8146	-54.5188
	5.00	2400.00000*	459.46829	.003	887.8521	3912.1479
5	1.00	-3300.00000*	459.46829	.000	-4812.1479	-1787.8521
	2.00	-4900.00000*	459.46829	.000	-6412.1479	-3387.8521
	3.00	-3966.66667*	459.46829	.000	-5478.8146	-2454.5188
	4.00	-2400.00000*	459.46829	.003	-3912.1479	-887.8521

*. The mean difference is significant at the 0.05 level.

T = Trial

Table 5.10 Post Hoc, Tukey HSD multiple comparison statistical analysis between independent variables in each medium trials and dependent variable total biogas volume (*C. paraputrificum*)

(I) Trial	(J) Trial	Mean Difference	Std. Error	Sig.	95% Confidence Interval	
					Lower Bound	Upper Bound
T1	T2	-933.33333*	217.73242	.010	-1664.6793	-201.9874
	T3	-2066.66667*	217.73242	.000	-2798.0126	-1335.3207
	T4	-3166.66667*	217.73242	.000	-3898.0126	-2435.3207
	T5	-2966.66667*	217.73242	.000	-3698.0126	-2235.3207
	T6	-466.66667	217.73242	.329	-1198.0126	264.6793
T2	T1	933.33333*	217.73242	.010	201.9874	1664.6793
	T3	-1133.33333*	217.73242	.002	-1864.6793	-401.9874
	T4	-2233.33333*	217.73242	.000	-2964.6793	-1501.9874
	T5	-2033.33333*	217.73242	.000	-2764.6793	-1301.9874
	T6	466.66667	217.73242	.329	-264.6793	1198.0126
T3	T1	2066.66667*	217.73242	.000	1335.3207	2798.0126
	T2	1133.33333*	217.73242	.002	401.9874	1864.6793
	T4	-1100.00000*	217.73242	.003	-1831.3460	-368.6540
	T5	-900.00000*	217.73242	.014	-1631.3460	-168.6540
	T6	1600.00000*	217.73242	.000	868.6540	2331.3460
T4	T1	3166.66667*	217.73242	.000	2435.3207	3898.0126
	T2	2233.33333*	217.73242	.000	1501.9874	2964.6793
	T3	1100.00000*	217.73242	.003	368.6540	1831.3460
	T5	200.00000	217.73242	.934	-531.3460	931.3460
	T6	2700.00000*	217.73242	.000	1968.6540	3431.3460
T5	T1	2966.66667*	217.73242	.000	2235.3207	3698.0126
	T2	2033.33333*	217.73242	.000	1301.9874	2764.6793
	T3	900.00000*	217.73242	.014	168.6540	1631.3460
	T4	-200.00000	217.73242	.934	-931.3460	531.3460
	T6	2500.00000*	217.73242	.000	1768.6540	3231.3460
T6	T1	466.66667	217.73242	.329	-264.6793	1198.0126
	T2	-466.66667	217.73242	.329	-1198.0126	264.6793
	T3	-1600.00000*	217.73242	.000	-2331.3460	-868.6540
	T4	-2700.00000*	217.73242	.000	-3431.3460	-1968.6540
	T5	-2500.00000*	217.73242	.000	-3231.3460	-1768.6540

*. The mean difference is significant at the 0.05 level.

Table 5.11 Post Hoc, Tukey HSD multiple comparison statistical analysis between independent variables initial pH in different trials and dependent Variable final culture pH (*C. beijerinckii*).

(I) Trial	(J) Trial	Mean Difference (I-J)	Std. Error	Sig.	95% Confidence Interval	
					Lower Bound	Upper Bound
1.00	2.00	.2267*	.02082	.000	.1567	.2966
	3.00	.5500*	.02082	.000	.4801	.6199
	4.00	.7167*	.02082	.000	.6467	.7866
	5.00	.8000*	.02082	.000	.7301	.8699
	6.00	.8200*	.02082	.000	.7501	.8899
2.00	1.00	-.2267*	.02082	.000	-.2966	-.1567
	3.00	.3233*	.02082	.000	.2534	.3933
	4.00	.4900*	.02082	.000	.4201	.5599
	5.00	.5733*	.02082	.000	.5034	.6433
	6.00	.5933*	.02082	.000	.5234	.6633
3.00	1.00	-.5500*	.02082	.000	-.6199	-.4801
	2.00	-.3233*	.02082	.000	-.3933	-.2534
	4.00	.1667*	.02082	.000	.0967	.2366
	5.00	.2500*	.02082	.000	.1801	.3199
	6.00	.2700*	.02082	.000	.2001	.3399
4.00	1.00	-.7167*	.02082	.000	-.7866	-.6467
	2.00	-.4900*	.02082	.000	-.5599	-.4201
	3.00	-.1667*	.02082	.000	-.2366	-.0967
	5.00	.0833*	.02082	.017	.0134	.1533
	6.00	.1033*	.02082	.003	.0334	.1733
5.00	1.00	-.8000*	.02082	.000	-.8699	-.7301
	2.00	-.5733*	.02082	.000	-.6433	-.5034
	3.00	-.2500*	.02082	.000	-.3199	-.1801
	4.00	-.0833*	.02082	.017	-.1533	-.0134
	6.00	.0200	.02082	.922	-.0499	.0899
6.00	1.00	-.8200*	.02082	.000	-.8899	-.7501
	2.00	-.5933*	.02082	.000	-.6633	-.5234
	3.00	-.2700*	.02082	.000	-.3399	-.2001
	4.00	-.1033*	.02082	.003	-.1733	-.0334
	5.00	-.0200	.02082	.922	-.0899	.0499

Based on observed means.

The error term is Mean Square(Error) = .001.

*. The mean difference is significant at the 0.05 level.

Table 5.12 Post Hoc, Tukey HSD multiple comparison statistical analysis between independent variables initial pH in different trials and dependent Variable final culture pH (*C. paraputrificum*).

(I) Trial	(J) Trial	Mean Difference (I-J)	Std. Error	Sig.	95% Confidence Interval	
					Lower Bound	Upper Bound
1.00	2.00	.2267*	.02082	.000	.1567	.2966
	3.00	.5500*	.02082	.000	.4801	.6199
	4.00	.7167*	.02082	.000	.6467	.7866
	5.00	.8000*	.02082	.000	.7301	.8699
	6.00	.8200*	.02082	.000	.7501	.8899
2.00	1.00	-.2267*	.02082	.000	-.2966	-.1567
	3.00	.3233*	.02082	.000	.2534	.3933
	4.00	.4900*	.02082	.000	.4201	.5599
	5.00	.5733*	.02082	.000	.5034	.6433
	6.00	.5933*	.02082	.000	.5234	.6633
3.00	1.00	-.5500*	.02082	.000	-.6199	-.4801
	2.00	-.3233*	.02082	.000	-.3933	-.2534
	4.00	.1667*	.02082	.000	.0967	.2366
	5.00	.2500*	.02082	.000	.1801	.3199
	6.00	.2700*	.02082	.000	.2001	.3399
4.00	1.00	-.7167*	.02082	.000	-.7866	-.6467
	2.00	-.4900*	.02082	.000	-.5599	-.4201
	3.00	-.1667*	.02082	.000	-.2366	-.0967
	5.00	.0833*	.02082	.017	.0134	.1533
	6.00	.1033*	.02082	.003	.0334	.1733
5.00	1.00	-.8000*	.02082	.000	-.8699	-.7301
	2.00	-.5733*	.02082	.000	-.6433	-.5034
	3.00	-.2500*	.02082	.000	-.3199	-.1801
	4.00	-.0833*	.02082	.017	-.1533	-.0134
	6.00	.0200	.02082	.922	-.0499	.0899
6.00	1.00	-.8200*	.02082	.000	-.8899	-.7501
	2.00	-.5933*	.02082	.000	-.6633	-.5234
	3.00	-.2700*	.02082	.000	-.3399	-.2001
	4.00	-.1033*	.02082	.003	-.1733	-.0334
	5.00	-.0200	.02082	.922	-.0899	.0499

Based on observed means.

The error term is Mean Square(Error) = .001.

*. The mean difference is significant at the 0.05 level.

Table 5.13 Post Hoc, Tukey HSD multiple comparison statistical analysis between independent variables initial biomass in different trials and dependent variable final biomass (*C. beijerinckii*).

(I) Trial	(J) Trial	Mean Difference (I-J)	Std. Error	Sig.	95% Confidence Interval	
					Lower Bound	Upper Bound
1.00	2.00	-.8230*	.08544	.000	-1.1042	-.5418
	3.00	-1.1535*	.08544	.000	-1.4347	-.8724
	4.00	-.4531*	.08544	.002	-.7343	-.1719
	5.00	.2504	.08544	.087	-.0308	.5316
2.00	1.00	.8230*	.08544	.000	.5418	1.1042
	3.00	-.3306*	.08544	.020	-.6117	-.0494
	4.00	.3699*	.08544	.010	.0887	.6511
	5.00	1.0734*	.08544	.000	.7922	1.3545
3.00	1.00	1.1535*	.08544	.000	.8724	1.4347
	2.00	.3306*	.08544	.020	.0494	.6117
	4.00	.7004*	.08544	.000	.4193	.9816
	5.00	1.4039*	.08544	.000	1.1227	1.6851
4.00	1.00	.4531*	.08544	.002	.1719	.7343
	2.00	-.3699*	.08544	.010	-.6511	-.0887
	3.00	-.7004*	.08544	.000	-.9816	-.4193
	5.00	.7035*	.08544	.000	.4223	.9846
5.00	1.00	-.2504	.08544	.087	-.5316	.0308
	2.00	-1.0734*	.08544	.000	-1.3545	-.7922
	3.00	-1.4039*	.08544	.000	-1.6851	-1.1227
	4.00	-.7035*	.08544	.000	-.9846	-.4223

Based on observed means.

The error term is Mean Square(Error) = .011.

*. The mean difference is significant at the 0.05 level.

Table 5.14 Student t test group statistics showing the mean concentrations of organic acids produced in all trials for *C. beijerinckii* cultures.

	Organic acids g/l	N	Mean	Std. Deviation	Std. Error Mean
g/l	Acetate	15	3.6149	1.31587	.33976
	Butyrate	15	3.2101	1.22623	.31661

g/l = grams per litre

N = Number of factors

Table 5.15 Student t test showing the difference between final acetate and butyrate concentrations in *C. beijerinckii* culture.

		Levene's Test for Equality of Variances	
		F	Sig.
g/l	Equal variances assumed	.036	.852
	Equal variances not assumed		

Acetate and butyrate		t-test for Equality of Means				
		t	df	Sig. (2-tailed)	Mean Difference	Std. Error Difference
g/l	Equal variances assumed	.872	28	.391	.40482	.46441
	Equal variances not assumed	.872	27.862	.391	.40482	.46441

		t-test for Equality of Means	
		95% Confidence Interval of the Difference	
		Lower	Upper
g/l	Equal variances assumed	-.54648	1.35613
	Equal variances not assumed	-.54669	1.35634

Table 5.16 Comparing the mean differences between acetate and butyrate in all trials for *C. beijerinckii*.

	Organic acids g/l	N	Mean	Std. Deviation	Std. Error Mean
T1	Acetate	3	3.8210	.02562	.01479
	Butyrate	3	3.4172	.02568	.01483
T2	Acetate	3	4.5102	.02496	.01441
	Butyrate	3	4.5919	.01906	.01101
T3	Acetate	3	5.0837	.02609	.01506
	Butyrate	3	3.6602	.71582	.41328
T4	Acetate	3	3.2718	.01486	.00858
	Butyrate	3	2.8079	.02587	.01493
T5	Acetate	3	1.3878	.00553	.00319
	Butyrate	3	1.1688	.08164	.04713

Table 5.17 Student t-test analysis showing the significant differences between acetate and butyrate concentrations in all trials in the *C. beijerinckii* cultures.

		t-test for Equality of Means				
		t	df	Sig. (2-tailed)	Mean Difference	Std. Error Difference
T1	Equal variances assumed	19.279	4	.000	.40380	.02095
	Equal variances not assumed	19.279	4.000	.000	.40380	.02095
T2	Equal variances assumed	-4.503	4	.011	-.08166	.01813
	Equal variances not assumed	-4.503	3.741	.013	-.08166	.01813
T3	Equal variances assumed	3.442	4	.026	1.42342	.41355
	Equal variances not assumed	3.442	2.005	.075	1.42342	.41355
T4	Equal variances assumed	26.934	4	.000	.46391	.01722
	Equal variances not assumed	26.934	3.191	.000	.46391	.01722
T5	Equal variances assumed	4.637	4	.010	.21904	.04724
	Equal variances not assumed	4.637	2.018	.043	.21904	.04724

Independent Samples Test

		t-test for Equality of Means	
		95% Confidence Interval of the Difference	
		Lower	Upper
T1	Equal variances assumed	.34565	.46196
	Equal variances not assumed	.34565	.46196
T2	Equal variances assumed	-.13200	-.03131
	Equal variances not assumed	-.13341	-.02990
T3	Equal variances assumed	.27521	2.57163
	Equal variances not assumed	-.35145	3.19829
T4	Equal variances assumed	.41609	.51173
	Equal variances not assumed	.41091	.51691
T5	Equal variances assumed	.08788	.35020
	Equal variances not assumed	.01755	.42054

Table 5.18 Comparing the significant differences between acetate, butyrate and lactate total concentrations in all trials in the *C. paraputrificum* cultures.

Organic acids	Organic acids	Mean Difference	Std. Error	Sig.	95% Confidence Interval	
					Lower Bound	Upper Bound
Lactate	Acetate	-3.06152*	.41162	.000	-4.0552	-2.0679
	Butyrate	.41383	.41162	.577	-.5798	1.4075
Acetate	Lactate	3.06152*	.41162	.000	2.0679	4.0552
	Butyrate	3.47535*	.41162	.000	2.4817	4.4690
Butyrate	Lactate	-.41383	.41162	.577	-1.4075	.5798
	Acetate	-3.47535*	.41162	.000	-4.4690	-2.4817

*. The mean difference is significant at the 0.05 level.

Table 5.19 Post Hoc, Tukey HSD multiple comparison statistical analysis showing statisticl differences between organic acids produced in different trials of *C. beijerinckii* cultures.

Dependent Variable	Organic acids	Organic acids	Mean Difference	Std. Error	Sig.
T1	Lactate	Acetate	-3.01714*	.08508	.000
		Butyrate	-1.03208*	.08508	.000
	Acetate	Lactate	3.01714*	.08508	.000
		Butyrate	1.98506*	.08508	.000
	Butyrate	Lactate	1.03208*	.08508	.000
		Acetate	-1.98506*	.08508	.000
T2	Lactate	Acetate	-2.38646*	.04627	.000
		Butyrate	.35691*	.04627	.001
	Acetate	Lactate	2.38646*	.04627	.000
		Butyrate	2.74336*	.04627	.000
	Butyrate	Lactate	-.35691*	.04627	.001
		Acetate	-2.74336*	.04627	.000
T3	Lactate	Acetate	-2.23935*	.04822	.000
		Butyrate	1.26348*	.04822	.000
	Acetate	Lactate	2.23935*	.04822	.000
		Butyrate	3.50283*	.04822	.000
	Butyrate	Lactate	-1.26348*	.04822	.000
		Acetate	-3.50283*	.04822	.000
T4	Lactate	Acetate	-3.57540*	.05864	.000
		Butyrate	.63155*	.05864	.000
	Acetate	Lactate	3.57540*	.05864	.000
		Butyrate	4.20695*	.05864	.000
	Butyrate	Lactate	-.63155*	.05864	.000
		Acetate	-4.20695*	.05864	.000
T5	Lactate	Acetate	-3.57540*	.05864	.000
		Butyrate	.63155*	.05864	.000
	Acetate	Lactate	3.57540*	.05864	.000
		Butyrate	4.20695*	.05864	.000
	Butyrate	Lactate	-.63155*	.05864	.000
		Acetate	-4.20695*	.05864	.000
T6	Lactate	Acetate	-3.57540*	.05864	.000
		Butyrate	.63155*	.05864	.000
	Acetate	Lactate	3.57540*	.05864	.000
		Butyrate	4.20695*	.05864	.000
	Butyrate	Lactate	-.63155*	.05864	.000
		Acetate	-4.20695*	.05864	.000

*. The mean difference is significant at the 0.05 level.

Table 5.20 Post Hoc, Tukey HSD multiple comparison statistical analysis showing statistical differences between organic acids produced in differenct trials of *C. beijerinckii* cultures.

Dependent Variable	Organic acids	Organic acids	95% Confidence Interval	
			Lower Bound	Upper Bound
T1	Lactate	Acetate	-3.2782	-2.7561
		Butyrate	-1.2931	-.7710
	Acetate	Lactate	2.7561	3.2782
		Butyrate	1.7240	2.2461
	Butyrate	Lactate	.7710	1.2931
		Acetate	-2.2461	-1.7240
T2	Lactate	Acetate	-2.5284	-2.2445
		Butyrate	.2149	.4989
	Acetate	Lactate	2.2445	2.5284
		Butyrate	2.6014	2.8853
	Butyrate	Lactate	-.4989	-.2149
		Acetate	-2.8853	-2.6014
T3	Lactate	Acetate	-2.3873	-2.0914
		Butyrate	1.1155	1.4114
	Acetate	Lactate	2.0914	2.3873
		Butyrate	3.3549	3.6508
	Butyrate	Lactate	-1.4114	-1.1155
		Acetate	-3.6508	-3.3549
T4	Lactate	Acetate	-3.7553	-3.3955
		Butyrate	.4516	.8115
	Acetate	Lactate	3.3955	3.7553
		Butyrate	4.0270	4.3869
	Butyrate	Lactate	-.8115	-.4516
		Acetate	-4.3869	-4.0270
T5	Lactate	Acetate	-3.7553	-3.3955
		Butyrate	.4516	.8115
	Acetate	Lactate	3.3955	3.7553
		Butyrate	4.0270	4.3869
	Butyrate	Lactate	-.8115	-.4516
		Acetate	-4.3869	-4.0270
T6	Lactate	Acetate	-3.7553	-3.3955
		Butyrate	.4516	.8115
	Acetate	Lactate	3.3955	3.7553
		Butyrate	4.0270	4.3869
	Butyrate	Lactate	-.8115	-.4516
		Acetate	-4.3869	-4.0270

Table 5.21 Correlation between total biogas volume produced and final acetate concentrations in all trials for *C. beijerinckii* culture.

		Biogas	Acetate
Biogas (ml/l)	Pearson Correlation	1	.913**
	Sig. (2-tailed)		.000
	N	15	15
Acetate (g/l)	Pearson Correlation	.913**	1
	Sig. (2-tailed)	.000	
	N	15	15

**. Correlation is significant at the 0.01 level (2-tailed).

Table 5.22 Correlation between total biogas volume produced and final butyrate concentrations in all trials for *C. beijerinckii* culture.

		Biogas	Butyrate
Biogas (ml/l)	Pearson Correlation	1	.961**
	Sig. (2-tailed)		.000
	N	15	15
Butyrate (g/l)	Pearson Correlation	.961**	1
	Sig. (2-tailed)	.000	
	N	15	15

**. Correlation is significant at the 0.01 level (2-tailed).

Table 5.23 Correlation between total biogas volume produced and final organic acid concentrations in all trials for *C. beijerinckii* culture.

		Biogas	Organic acids
Biogas (ml/l)	Pearson Correlation	1	.946**
	Sig. (2-tailed)		.000
	N	15	15
Organic acids (g/l)	Pearson Correlation	.946**	1
	Sig. (2-tailed)	.000	
	N	15	15

**. Correlation is significant at the 0.01 level (2-tailed).

Table 5.24 Correlation between total biogas volume produced and final acetate concentrations in trials 3, 4 and 5 for *C. beijerinckii* culture.

		Biogas	Acetate
Biogas	Pearson Correlation	1	.941[**]
	Sig. (2-tailed)		.000
	N	9	9
Acetate	Pearson Correlation	.941[**]	1
	Sig. (2-tailed)	.000	
	N	9	9

**. Correlation is significant at the 0.01 level (2-tailed).

Table 5.25 Correlation between total biogas volume produced and final butyrate concentrations in trials 3, 4 and 5 for *C. beijerinckii* culture.

		Biogas	Butyrate
Biogas	Pearson Correlation	1	.945[**]
	Sig. (2-tailed)		.000
	N	9	9
Butyrate	Pearson Correlation	.945[**]	1
	Sig. (2-tailed)	.000	
	N	9	9

**. Correlation is significant at the 0.01 level (2-tailed).

Table 5.26 Correlation between total biogas volume produced and final organic acid concentrations in trials 3, 4 and 5 for *C. beijerinckii* culture

		Biogas	Organic acids
Biogas	Pearson Correlation	1	.943[**]
	Sig. (2-tailed)		.000
	N	9	9
Organic acids	Pearson Correlation	.943[**]	1
	Sig. (2-tailed)	.000	
	N	9	9

Table 5.27 Yield coefficients of products, acetate to butyrate ratios and yield of CO_2 and hydrogen produced in the *C. beijerinckii* culture.

Yield coefficient of product (mol of product/mol GlcNAc utilised)		Acetate/ Butyrate (mol/mol)	Biogas Volume (ml/l)	Moles of CO_2 expelled/ mol GlcNAc	Moles of H_2 expelled / mol GlcNAc
Acetate	Butyrate				
1.27	0.77	1.65	4133	0.30	2.90
1.07	0.74	1.45	5733	0.36	2.79
1.13	0.61	1.85	4800	0.41	2.08
1.18	0.69	1.72	3233	0.36	2.34
1.44	0.82	1.75	666	0.23	1.37

Table 5.28 Yield coefficients of products, acetate to butyrate ratios and yield of CO_2 and hydrogen produced in the *C. paraputrificum* culture.

Yield coefficient of product (mol of product/mol of GlcNAc utilised)			Acetate/ Butyrate (mol/mol)	Biogas Volume (ml/l)	Moles of CO_2 expelled/ mol GlcNAc	Moles of H_2 expelled / mol GlcNAc
Lactate	Acetate	Butyrate				
0.21	1.40	0.47	2.99	2366	0.34	1.62
0.46	1.32	0.41	3.25	3300	0.29	1.73
0.54	1.25	0.38	3.27	4433	0.29	1.73
0.47	1.30	0.40	3.21	5400	0.30	1.79
0.51	1.23	0.38	3.23	5333	0.31	1.58
0.46	1.52	0.53	2.87	2833	0.18	1.86

Table 5.29 Correlations between acetate to butyrate ratio in relation to hydrogen yield per mole of N-acetylglucosamine for *C. paraputrificum* cultures.

		Acetate / Butyrate ratio	Moles H_2/mol GlcNAG
Acetate / Butyrate ratio	Pearson Correlation	1	-.326
	Sig. (2-tailed)		.529
	N	6	6
Moles H_2/mol GlcNAG	Pearson Correlation	-.326	1
	Sig. (2-tailed)	.529	
	N	6	6

Table 5.30 Correlations between acetate to butyrate ratio in relation to hydrogen yield per mole of N-acetylglucosamine for *C. beijerinckii* cultures.

		Acetate / Butyrate ratio	Moles H$_2$/mol GlcNAG
Acetate / Butyrate ratio	Pearson Correlation	1	.143
	Sig. (2-tailed)		.819
	N	5	5
Moles H$_2$/mol GlcNAG	Pearson Correlation	.143	1
	Sig. (2-tailed)	.819	
	N	5	5

Appendix 6

Appendix 6.1 Box-Behnken culture designs for *C. beijerinckii* and *C. paraputrificum* cultures.

Table 6.1 Total biogas volumes in triplicate for *C. beijerinckii* cultures with N-acetylglucosamine as the main carbon source using the Box-Behnken design.

Trial Number	Biogas Volume (ml)			Biogas Volume Average (ml)	S.D	S.E
	a	b	c			
1	48	47	42	45.67	26.85	15.50
2	65	65	61	63.67	36.37	21.00
3	47	37	42	42.00	23.07	13.32
4	0	0	63	63.00	0.00	0.00
5	36	36	44	38.67	17.90	10.33
6	56	60	66	60.67	30.09	17.37
7	33	42	0	37.50	18.38	13.00
8	0	64	42	53.00	45.25	26.13
9	52	52	44	49.33	24.83	14.33
10	52	0	52	52.00	0.00	0.00
11	0	0	58	58.00	0.00	0.00
12	52	52	49	51.00	23.09	13.33
13	50	52	50	50.67	21.96	12.68
14	50	52	51	51.00	21.39	12.35
15	53	51	49	51.00	21.39	12.35

ml = millilitres.
S.D = Standard deviation.
S.E = Standard error.

Table 6.2 Biomass concentrations in triplicate for *C. beijerinckii* cultures with N-acetylglucosamine as the main carbon source using the Box-Behnken design.

Trial Number	Final O.D a	b	c	O.D Average	S.D	S.E
1	3.435	3.275	3.375	3.362	0.08	0.06
2	4.000	3.815	3.830	3.882	0.10	0.07
3	3.115	2.945	2.390	2.817	0.38	0.27
4	3.650	3.690	3.670	3.670	0.02	0.01
5	3.480	3.275	3.405	3.387	0.10	0.07
6	3.480	3.795	3.345	3.540	0.23	0.16
7	1.975	2.145	2.145	2.060	0.10	0.07
8	2.706	3.215	2.705	2.960	0.29	0.21
9	3.485	3.540	3.470	3.498	0.04	0.03
10	3.625	2.585	3.510	3.568	0.57	0.40
11	2.485	2.685	2.585	2.585	0.10	0.07
12	3.575	2.850	3.070	3.165	0.37	0.26
13	3.500	3.630	2.255	3.128	0.76	0.54
14	3.765	3.510	3.390	3.555	0.19	0.14
15	3.765	3.195	3.670	3.543	0.31	0.22

O.D = Culture optical density in absorbance unit.
S.D = Standard deviation.
S.E = Standard error.

Table 6.3 Culture final pH in triplicate for *C. beijerinckii* cultures with N-acetylglucosamine as the main carbon source using the Box-Behnken design.

Culture	Final pH			Final pH		
Trials	a	b	c	Average	S.D	S.E
1	5.54	5.53	5.51	5.53	0.02	0.01
2	5.45	5.47	5.47	5.46	0.01	0.01
3	5.50	5.48	5.54	5.51	0.03	0.02
4	5.50	5.50	5.50	5.50	0.00	0.00
5	5.33	5.33	5.33	5.33	0.00	0.00
6	5.25	5.25	5.28	5.26	0.02	0.01
7	5.76	5.78	5.76	5.77	0.01	0.01
8	5.68	5.65	5.67	5.66	0.02	0.01
9	5.31	5.30	5.33	5.31	0.02	0.01
10	5.26	5.30	5.31	5.29	0.03	0.02
11	5.74	5.74	5.74	5.74	0.00	0.00
12	5.71	5.68	5.67	5.69	0.02	0.01
13	5.43	5.40	5.43	5.42	0.02	0.01
14	5.44	5.46	5.44	5.45	0.01	0.01
15	5.47	5.48	5.46	5.47	0.01	0.01

S.D = Standard deviation.
S.E = Standard error.

Table 6.4 Total biogas volume in triplicate for *C. paraputrificum* cultures with N-acetylglucosamine as the main carbon source using the Box-Behnken design.

Trial Number	Biogas Volume (ml) a	b	c	Biogas Volume Average (ml)	S.D	S.E
1	39	42	42	41.00	1.73	1.00
2	51	46	44	47.00	3.61	2.08
3	35	33	35	34.33	1.15	0.67
4	47	48	52	49.00	2.65	1.53
5	49	46	46	47.00	1.73	1.00
6	52	52	51	51.67	0.58	0.33
7	42	42	45	43.00	1.73	1.00
8	51	46	49	48.67	2.52	1.45
9	51	52	52	51.67	0.58	0.33
10	46	51	51	49.33	2.89	1.67
11	48	51	50	49.67	1.53	0.88
12	43	35	48	42.00	6.56	3.79
13	49	49	49	49.00	0.00	0.00
14	48	47	52	49.00	2.65	1.53
15	51	51	44	48.67	4.04	2.33

ml = millilitres.
S.D = Standard deviation.
S.E = Standard error.

Table 6.5 Biomass concentration in triplicate for *C. paraputrificum* cultures with N-acetylglucosamine as the main carbon source using the Box-Behnken design.

Trial Number	Final O.D			O.D Average	S.D	S.E
	a	b	c			
1	1.623	1.584	1.404	1.537	0.12	0.07
2	1.625	1.773	1.575	1.658	0.10	0.06
3	1.542	1.422	1.566	1.510	0.08	0.04
4	1.497	1.326	1.140	1.321	0.18	0.10
5	1.113	0.735	1.038	0.962	0.20	0.12
6	1.551	1.476	1.382	1.470	0.08	0.05
7	1.470	1.383	1.269	1.374	0.10	0.06
8	1.559	1.506	1.653	1.573	0.07	0.04
9	1.641	1.857	1.605	1.701	0.14	0.08
10	1.266	1.014	1.224	1.168	0.14	0.08
11	1.521	1.497	1.251	1.423	0.15	0.09
12	1.134	1.233	1.131	1.166	0.06	0.03
13	1.266	1.350	1.353	1.323	0.05	0.03
14	1.344	1.455	1.530	1.443	0.09	0.05
15	1.575	1.365	1.248	1.396	0.17	0.10

O.D = Culture optical density in absorbance units.
S.D = Standard deviation.
S.E = Standard error.

Table 6.6 Culture final pH in triplicate for *C. paraputrificum* cultures with N-acetylglucosamine as the main carbon source using the Box-Behnken design.

Culture Trials	Final pH a	b	c	Final pH Average	S.D	S.E
1	5.07	5.13	5.10	5.10	0.03	0.02
2	5.00	4.99	4.99	4.99	0.01	0.00
3	5.21	5.18	5.20	5.20	0.02	0.01
4	5.00	5.01	5.07	5.03	0.04	0.02
5	5.16	5.18	5.17	5.17	0.01	0.01
6	5.04	5.01	5.02	5.02	0.02	0.01
7	5.08	5.11	5.12	5.10	0.02	0.01
8	5.01	4.97	4.99	4.99	0.02	0.01
9	5.09	5.10	5.13	5.11	0.02	0.01
10	5.13	5.12	5.12	5.12	0.01	0.00
11	5.11	5.11	5.15	5.12	0.02	0.01
12	5.20	5.14	5.19	5.18	0.03	0.02
13	5.11	5.06	5.08	5.08	0.03	0.01
14	5.11	5.09	5.14	5.11	0.03	0.01
15	5.12	5.09	5.12	5.11	0.02	0.01

S.D = Standard deviation.
S.E = Standard error.

Table 6.7 Box-Behnken design with three independent variables used by *C. beijerinckii* cultures.

Trial Number	N-acetylglucosamine (g/l) X_1	Code X_1	FeSO$_4$.7H$_2$0 (g/l) X_2	Code X_2	Initial pH X_3	Code X_3	Biogas Volume (ml/l)
1	13	-1	0.06	-1	6.3	0	4567
2	21	1	0.06	-1	6.3	0	6367
3	13	-1	0.10	1	6.3	0	4200
4	21	1	0.10	1	6.3	0	6300
5	13	-1	0.08	0	6.1	-1	3867
6	21	1	0.08	0	6.1	-1	6067
7	13	-1	0.08	0	6.5	1	3750
8	21	1	0.08	0	6.5	1	5300
9	17	0	0.06	-1	6.1	-1	4933
10	17	0	0.10	1	6.1	-1	5200
11	17	0	0.06	-1	6.5	1	5800
12	17	0	0.10	1	6.5	1	5100
13	17	0	0.08	0	6.3	0	5067
14	17	0	0.08	0	6.3	0	5100
15	17	0	0.08	0	6.3	0	5100

X₁ = N-acetylglucosamine.

X_1 = N-acetylglucosamine.
X_2 = FeSO4.7H₂O.
X_3 = Initial pH.
g/l = grams per litre
ml/l = millilitres per litre.

Table 6.8 Box-Behnken design with three independent variables used by *C. paraputrificum* cultures.

Trial Number	N-acetylglucosamine (g/l)		L-cysteine.HCl (g/l)		MgCl₂ (g/l)		Biogas Volume (ml/l)
	X_1	Code X_1	X_2	Code X_2	X_3	Code X_3	
1	23	-1	0.10	-1	0.45	0	4100
2	31	1	0.10	-1	0.45	0	4700
3	23	-1	0.40	1	0.45	0	3433
4	31	1	0.40	1	0.45	0	4900
5	23	-1	0.25	0	0.40	-1	4700
6	31	1	0.25	0	0.40	-1	5166
7	23	-1	0.25	0	0.50	1	4300
8	31	1	0.25	0	0.50	1	4866
9	27	0	0.10	-1	0.40	-1	5166
10	27	0	0.40	1	0.40	-1	4933
11	27	0	0.10	-1	0.50	1	4966
12	27	0	0.40	1	0.50	1	4200
13	27	0	0.25	0	0.45	0	4900
14	27	0	0.25	0	0.45	0	4900
15	27	0	0.25	0	0.45	0	4866

X_1 = N-acetylglucosamine.
X_2 = L-cysteine.HCl.H₂O.
X_3 = MgCl₂.
g/l = grams per litre
ml/l = millilitres per litre.

Table 6.9 total biogas volume for *C. beijerinckii* and *C. paraputrificum* cultures in triplicate using the optimized statistical set conditions.

Clostridium Species	Biogas Volume (ml) a	b	c	d	Biogas Average (ml)	S.D	S.E
Clostridium beijerinckii	67	65	64	-	65.33	1.53	0.88
Clostridium paraputrificum	53	53	54	54	53.50	0.58	0.29

ml = millilitres.
S.D = Standard deviation.
S.E = Standard error.

Table 6.10 Biomass concentration for *C. beijerinckii* and *C. paraputrificum* cultures in triplicate using the optimized statistical set conditions.

Clostridium Species	Final O.D a	b	c	d	O.D Average (AU)	S.D	S.E
Clostridium beijerinckii	4.4.25	4.569	4.420	-	4.495	0.11	0.06
Clostridium paraputrificum	3.200	2.901	2.981	3.017	3.025	0.13	0.03

O.D = Culture optical density in absorbance unit.
S.D = Standard deviation.
S.E = Standard error.

Table 6.11 Final culture pH for *C. beijerinckii* and *C. paraputrificum* cultures in triplicate using the optimized statistical set conditions.

Clostridium Species	Final pH a	b	c	d	Final pH Average	S.D	S.E
Clostridium beijerinckii	5.24	5.23	5.23	-	5.23	0.01	0.00
Clostridium paraputrificum	5.11	5.07	5.08	5.09	5.09	0.02	0.01

S.D = Standard deviation.
S.E = Standard error.

Table 6.12 Comparing the mean differences between acetate and butyrate in all trials for *C. beijerinckii* cultures

	Organic acids	N	Mean	Std. Deviation	Std. Error Mean
Organic acids (g/l)	Acetate	45	4.7072	.79319	.11824
	Butyrate	45	4.1812	.58571	.08731

Table 6.13 Independent t-test for equality of means showing the significant differences between organic acids produced in the *C. beijerinckii* cultures.

		Levene's Test for Equality of Variances		t-test for Equality of Means	
		F	Sig.	t	df
Organic acids (g/l)	Equal variances assumed	3.782	.055	3.579	88
	Equal variances not assumed			3.579	80.987

		t-test for Equality of Means		
		Sig. (2-tailed)	Mean Difference	Std. Error Difference
Organic acids (g/l)	Equal variances assumed	.001	.52599	.14698
	Equal variances not assumed	.001	.52599	.14698

		t-test for Equality of Means	
		95% Confidence Interval of the Difference	
		Lower	Upper
Organic acids (g/l)	Equal variances assumed	.23388	.81809
	Equal variances not assumed	.23353	.81844

Table 6.13 Analysis of variance test showing the overall significance between group trials and within group trials in the *C. paraputrificum* cultures.

	Sum of Squares	df	Mean Square	F	Sig.
Between Groups	1032.928	4	258.232	1825.230	.000
Within Groups	31.125	220	.141		
Total	1064.054	224			

Table 6.14 Post Hoc, Tukey HSD multiple comparison statistical analysis showing statistical differences between organic acids produced in differenct trials of *C. paraputrificum* cultures using the Box-behnken method.

(I) Organic acids	(J) Organic acids	Mean Difference (I-J)	Std. Error	Sig.	95% Confidence Interval	
					Lower Bound	Upper Bound
Lactate	Formate	2.29186*	.07930	.000	2.0738	2.5100
	Acetate	-3.41011*	.07930	.000	-3.6282	-3.1920
	Ethanol	2.48605*	.07930	.000	2.2679	2.7042
	Butyrate	-.32865*	.07930	.000	-.5468	-.1105
Formate	Lactate	-2.29186*	.07930	.000	-2.5100	-2.0738
	Acetate	-5.70197*	.07930	.000	-5.9201	-5.4839
	Ethanol	.19419	.07930	.106	-.0239	.4123
	Butyrate	-2.62051*	.07930	.000	-2.8386	-2.4024
Acetate	Lactate	3.41011*	.07930	.000	3.1920	3.6282
	Formate	5.70197*	.07930	.000	5.4839	5.9201
	Ethanol	5.89616*	.07930	.000	5.6781	6.1143
	Butyrate	3.08146*	.07930	.000	2.8634	3.2996
Ethanol	Lactate	-2.48605*	.07930	.000	-2.7042	-2.2679
	Formate	-.19419	.07930	.106	-.4123	.0239
	Acetate	-5.89616*	.07930	.000	-6.1143	-5.6781
	Butyrate	-2.81470*	.07930	.000	-3.0328	-2.5966
Butyrate	Lactate	.32865*	.07930	.000	.1105	.5468
	Formate	2.62051*	.07930	.000	2.4024	2.8386
	Acetate	-3.08146*	.07930	.000	-3.2996	-2.8634
	Ethanol	2.81470*	.07930	.000	2.5966	3.0328

*. The mean difference is significant at the 0.05 level.

Table 6.15 Comparing the mean differences between organic acids produced in all trials for *C. paraputrificum* cultures in the Box-behnken method.

Organic acids	Mean	N	Std. Deviation	Std. Error of Mean
Lactate	2.5520	45	.56569	.08433
Formate	.2602	45	.14531	.02166
Acetate	5.9621	45	.40618	.06055
Ethanol	.0660	45	.17153	.02557
Butyrate	2.8807	45	.41456	.06180
Total	2.3442	225	2.17950	.14530

N = total number of batch runs.

Table 6.16 Comparing the mean differences at different time points between organic acids produced in batch 1 liter bioreactor run for *C. beijerinckii*.

Time (h)	Organic acids	N	Mean	Std. Deviation	Std. Error Mean
0	Acetate	3	.0000	.00000[a]	.00000
	Butyrate	3	.0000	.00000[a]	.00000
4	Acetate	3	.0000	.00000[a]	.00000
	Butyrate	3	.0000	.00000[a]	.00000
21	Acetate	3	2.6442	.04114	.02375
	Butyrate	3	3.7818	.17319	.09999
25	Acetate	3	3.0334	.00517	.00299
	Butyrate	3	4.4951	.01047	.00605
28	Acetate	3	3.2189	.06964	.04021
	Butyrate	3	4.6446	.09907	.05720
31	Acetate	3	3.4931	.01253	.00723
	Butyrate	3	4.9291	.03873	.02236
49	Acetate	3	4.2608	.04356	.02515
	Butyrate	3	6.0462	.02792	.01612
73	Acetate	3	4.3336	.04786	.02763
	Butyrate	3	6.0022	.03274	.01890
75	Acetate	3	4.3720	.02204	.01272
	Butyrate	3	6.0099	.06702	.03869

a. t cannot be computed because the standard deviations of both groups are 0.

Table 6.17 Independent t-test for equality of means showing the significant differences between organic acids produced at each time points in the *C. beijerinckii* batch 1 liter cultures.

Time (h)		Levene's Test for Equality of Variances		t-test for Equality of Means	
		F	Sig.	t	df
21	Equal variances assumed	2.201	.212	-11.069	4
	Equal variances not assumed			-11.069	2.225
25	Equal variances assumed	.824	.415	-216.727	4
	Equal variances not assumed			-216.727	2.921
28	Equal variances assumed	.236	.652	-20.390	4
	Equal variances not assumed			-20.390	3.589
31	Equal variances assumed	1.657	.267	-61.105	4
	Equal variances not assumed			-61.105	2.414
49	Equal variances assumed	.365	.578	-59.764	4
	Equal variances not assumed			-59.764	3.406
73	Equal variances assumed	.272	.630	-49.843	4
	Equal variances not assumed			-49.843	3.536
75	Equal variances assumed	1.626	.271	-40.213	4
	Equal variances not assumed			-40.213	2.428

Time (h)		t-test for Equality of Means		
		Sig. (2-tailed)	Mean Difference	Std. Error Difference
21	Equal variances assumed	.000	-1.13760	.10278
	Equal variances not assumed	.005	-1.13760	.10278
25	Equal variances assumed	.000	-1.46172	.00674
	Equal variances not assumed	.000	-1.46172	.00674
28	Equal variances assumed	.000	-1.42563	.06992
	Equal variances not assumed	.000	-1.42563	.06992
31	Equal variances assumed	.000	-1.43595	.02350
	Equal variances not assumed	.000	-1.43595	.02350
49	Equal variances assumed	.000	-1.78531	.02987

	Equal variances not assumed	.000	-1.78531	.02987
73	Equal variances assumed	.000	-1.66857	.03348
	Equal variances not assumed	.000	-1.66857	.03348
75	Equal variances assumed	.000	-1.63789	.04073
	Equal variances not assumed	.000	-1.63789	.04073

Time (h)		t-test for Equality of Means	
		95% Confidence Interval of the Difference	
		Lower	Upper
21	Equal variances assumed	-1.42295	-.85225
	Equal variances not assumed	-1.53961	-.73559
25	Equal variances assumed	-1.48045	-1.44300
	Equal variances not assumed	-1.48352	-1.43993
28	Equal variances assumed	-1.61976	-1.23151
	Equal variances not assumed	-1.62886	-1.22241
31	Equal variances assumed	-1.50119	-1.37070
	Equal variances not assumed	-1.52213	-1.34976
49	Equal variances assumed	-1.86825	-1.70237
	Equal variances not assumed	-1.87428	-1.69635
73	Equal variances assumed	-1.76152	-1.57563
	Equal variances not assumed	-1.76654	-1.57061
75	Equal variances assumed	-1.75098	-1.52481
	Equal variances not assumed	-1.78664	-1.48914

Table 6.18 Comparing the overall mean differences between organic acids produced in batch 1 liter bioreactor run for *C. beijerinckii*.

	Organic acids	N	Mean	Std. Deviation	Std. Error Mean
Organic acids (g/l)	ACETATE	27	2.8173	1.64236	.31607
	BUTYRATE	27	3.9899	2.29977	.44259

Table 6.19 Independent t-test for equality of means showing the cumulative significant differences between organic acids produced in the *C. beijerinckii* batch 1 liter cultures.

		Levene's Test for Equality of Variances		t-test for Equality of Means	
		F	Sig.	t	df
Organic acids	Equal variances assumed	2.680	.108	-2.156	52
	Equal variances not assumed			-2.156	47.046

		t-test for Equality of Means		
		Sig. (2-tailed)	Mean Difference	Std. Error Difference
Organic acids	Equal variances assumed	.036	-1.17252	.54386
	Equal variances not assumed	.036	-1.17252	.54386

		t-test for Equality of Means	
		95% Confidence Interval of the Difference	
		Lower	Upper
Organic acids	Equal variances assumed	-2.26386	-.08118
	Equal variances not assumed	-2.26661	-.07843

Table 6.20 Analysis of variance showing the statistical differences between organic acids at each time points for *C. beijerinckii* batch cultures.

Time (h)		Sig.
0	Between Groups	.
	Within Groups	
	Total	
2	Between Groups	.
	Within Groups	
	Total	
6	Between Groups	.
	Within Groups	
	Total	
8	Between Groups	.
	Within Groups	
	Total	
26	Between Groups	.000
	Within Groups	
	Total	
30	Between Groups	.000
	Within Groups	
	Total	
35	Between Groups	.000
	Within Groups	
	Total	
38	Between Groups	.000
	Within Groups	
	Total	
48	Between Groups	.000
	Within Groups	
	Total	
58	Between Groups	.000
	Within Groups	
	Total	

Table 6.21 Post Hoc, Tukey HSD multiple comparison statistical analysis showing statistical differences between organic acids produced at different time points of batch 1 liter bioreactor run for *C. paraputrificum* cultures.

Dependent Variable Time (h)	(I) Organic acids	(J) Organic acids	Mean Difference (I-J)	Std. Error	Sig.
26	LACTATE	ACETAE	-2.42071*	.01438	.000
		BUTYRATE	-.32321*	.01438	.000
	ACETAE	LACTATE	2.42071*	.01438	.000
		BUTYRATE	2.09751*	.01438	.000
	BUTYRATE	LACTATE	.32321*	.01438	.000
		ACETAE	-2.09751*	.01438	.000
30	LACTATE	ACETAE	-2.56187*	.02836	.000
		BUTYRATE	-.23807*	.02836	.000
	ACETAE	LACTATE	2.56187*	.02836	.000
		BUTYRATE	2.32380*	.02836	.000
	BUTYRATE	LACTATE	.23807*	.02836	.000
		ACETAE	-2.32380*	.02836	.000
35	LACTATE	ACETAE	-2.63146*	.02553	.000
		BUTYRATE	-.01194	.02553	.888
	ACETAE	LACTATE	2.63146*	.02553	.000
		BUTYRATE	2.61951*	.02553	.000
	BUTYRATE	LACTATE	.01194	.02553	.888
		ACETAE	-2.61951*	.02553	.000
38	LACTATE	ACETAE	-2.55262*	.01132	.000
		BUTYRATE	-.01947	.01132	.273
	ACETAE	LACTATE	2.55262*	.01132	.000
		BUTYRATE	2.53314*	.01132	.000
	BUTYRATE	LACTATE	.01947	.01132	.273
		ACETAE	-2.53314*	.01132	.000
48	LACTATE	ACETAE	-2.54834*	.01417	.000
		BUTYRATE	.00337	.01417	.969
	ACETAE	LACTATE	2.54834*	.01417	.000
		BUTYRATE	2.55171*	.01417	.000
	BUTYRATE	LACTATE	-.00337	.01417	.969
		ACETAE	-2.55171*	.01417	.000
58	LACTATE	ACETAE	-2.55832*	.02250	.000
		BUTYRATE	-.04370	.02250	.208
	ACETAE	LACTATE	2.55832*	.02250	.000

Dependent Variable Time (h)	(I) Organic acids	(J) Organic acids	95% Confidence Interval	
			Lower Bound	Upper Bound
26	LACTATE	ACETAE	-2.4648*	-2.3766
		BUTYRATE	-.3673*	-.2791
	ACETAE	LACTATE	2.3766*	2.4648
		BUTYRATE	2.0534*	2.1416
	BUTYRATE	LACTATE	.2791*	.3673
		ACETAE	-2.1416*	-2.0534
30	LACTATE	ACETAE	-2.6489*	-2.4749
		BUTYRATE	-.3251*	-.1511
	ACETAE	LACTATE	2.4749*	2.6489
		BUTYRATE	2.2368*	2.4108
	BUTYRATE	LACTATE	.1511*	.3251
		ACETAE	-2.4108*	-2.2368
35	LACTATE	ACETAE	-2.7098*	-2.5531
		BUTYRATE	-.0903	.0664
	ACETAE	LACTATE	2.5531*	2.7098
		BUTYRATE	2.5412*	2.6978
	BUTYRATE	LACTATE	-.0664	.0903
		ACETAE	-2.6978*	-2.5412
38	LACTATE	ACETAE	-2.5874*	-2.5179
		BUTYRATE	-.0542	.0153
	ACETAE	LACTATE	2.5179*	2.5874
		BUTYRATE	2.4984*	2.5679
	BUTYRATE	LACTATE	-.0153	.0542
		ACETAE	-2.5679*	-2.4984
40	LACTATE	ACETAE	-2.5918*	-2.5049
		BUTYRATE	-.0401	.0468
	ACETAE	LACTATE	2.5049*	2.5918
		BUTYRATE	2.5083*	2.5952
	BUTYRATE	LACTATE	-.0468	.0401
		ACETAE	-2.5952*	-2.5083
58	LACTATE	ACETAE	-2.6274*	-2.4893
		BUTYRATE	-.1128	.0253
	ACETAE	LACTATE	2.4893*	2.6274

Dependent Variable Time (h)	(I) Organic acids	(J) organic acids	Mean Difference (I-J)	Std. Error	Sig.

353

58	ACETAE	BUTYRATE	2.51462*	.02250	.000
		LACTATE	.04370*	.02250	.208
	BUTYRATE	ACETAE	-2.51462*	.02250	.000

Dependent Variable Time (h)	(I) Organic acids	(J) Organic acids	95% Confidence Interval	
			Lower Bound	Upper Bound
58	ACETAE	BUTYRATE	2.4456*	2.5837
		LACTATE	-.0253*	.1128
	BUTYRATE	ACETAE	-2.5837*	-2.4456

*. The mean difference is significant at the 0.05 level.

Table 6.22 Post Hoc, Tukey HSD multiple comparison statistical analysis showing statistical differences between cummulative organic acids produced in batch 1 liter bioreactor run for *C. paraputrificum* cultures.

Organic acids (I)	Organic acids (J)	Mean Difference (I-J)	Std. Error	Sig.
LACTATE	ACETATE	-1.52733*	.48676	.007
	BUTYRATE	-.06330	.48676	.991
ACETATE	LACTATE	1.52733*	.48676	.007
	BUTYRATE	1.46403*	.48676	.010
BUTYRATE	LACTATE	.06330	.48676	.991
	ACETATE	-1.46403*	.48676	.010

Organic acids (I)	Organic acids (J)	95% Confidence Interval	
		Lower Bound	Upper Bound
LACTATE	ACETATE	-2.6880*	-.3667
	BUTYRATE	-1.2240	1.0974
ACETATE	LACTATE	.3667*	2.6880
	BUTYRATE	.3034*	2.6247
BUTYRATE	LACTATE	-1.0974	1.2240
	ACETATE	-2.6247*	-.3034

*. The mean difference is significant at the 0.05 level.

www.ingramcontent.com/pod-product-compliance
Lightning Source LLC
Chambersburg PA
CBHW080757180526
45168CB00006B/2241